The Dawn of Science

Thanu Padmanabhan · Vasanthi Padmanabhan

The Dawn of Science

Glimpses from History for the Curious Mind

Thanu Padmanabhan
Pune University Campus
IUCAA
Pune, Maharashtra, India

Vasanthi Padmanabhan
Pune University Campus
IUCAA
Pune, Maharashtra, India

ISBN 978-3-030-17508-5 ISBN 978-3-030-17509-2 (eBook)
https://doi.org/10.1007/978-3-030-17509-2

This Springer imprint is published by the registered company Springer Nature Switzerland AG
The registered company address is: Gewerbestrasse 11, 6330 Cham, Switzerland

Preface

The development of scientific ideas is probably one of the highest intellectual achievements of *Homo sapiens*. This arose from the efforts — and, as we will see throughout in this book, sacrifices — of a number of dedicated souls. Many of these could probably have quite successfully pursued totally different vocations in life! But they involved themselves with science because they were, by and large, driven by curiosity and were fascinated by the intricacies of Nature. This led them to devote — quite often — their entire lifetimes to unravelling the mysteries of the world around them.

The purpose of this book is to share the excitement the authors feel about the historical development of scientific ideas with the like-minded, curious, educated lay-public. Needless to say, this is *not* a monograph on the history of science written by a couple of historians of science for fellow historians of science. For one, such volumes — while no doubt, excellent in erudition — can be somewhat soporific in style; our intention is that you should actually *enjoy* reading this book!

To do this in a modular and entertaining fashion, we have just picked 24 topics covering different milestones in science from antiquity till about the 17th century. We have stopped with Newton's contribution to physics and Lavoisier's to chemistry, which we think could be thought of as the end of a period that we would call the *dawn* of science. This end point, as well as the choice of the 24 milestones, does of course reflect our personal preferences. But we are quite sure that at least 20 out of these 24 topics would feature in any sensible listing of the early milestones in the history of science. All these chapters (except possibly one or two) will be accessible to anyone with an exposure to science at the high-school level.

There are some unique features which we would like to alert you to, and which should help you to enjoy it more:

- To begin with, the book is completely modular and you can dip into any of the chapters independently of the others; so you can read the chapters in any order that appeals to you. The development of ideas is presented, by and large, chronologically and you might appreciate some chapters more if you read them along with closely related, adjacent, chapters — but this is by no means mandatory. The modularity, which makes each chapter reasonably self-contained, also implies that the book should be viewed more like a collection of essays on a connecting theme, rather than as a conventional book made up of several chapters.

- Second, you can also enjoy the material presented in the "boxes", most of the time, without actually reading through the main text in the chapter. This is done, once again, in the spirit of modularity, and if the boxes in a particular chapter attract your attention, then you may like to read it in full! The same goes for the figures and figure captions. We have tried to make history come alive by providing a fair number of figures — typically half a dozen per chapter — with reasonably self-contained figure captions. You may want to glance through them to get an initial flavour of what the chapter contains. The figure captions include a certain amount of unavoidable text overlap with the main body of the chapter, but this should be considered a feature rather than a bug!

- Third, all the chapters contain a special diagram entitled "When" (and some of the chapters also contain a diagram entitled "Where"). This diagram summarizes the events in the historical period described in that particular chapter. To the left of the timeline, you can see the key events in science, while to the right of the timeline we have given the key events in world history (and Indian history, as you would naturally expect from two Indian authors!). This will allow you to appreciate the way science interfaced with social structures and political events as it developed.

- Fourth, we have interpreted the development of scientific ideas in a reasonably broad context. For example, we have included chapters on the exploration of the high seas, the story of the calendar system, and the development of printing in this collection of milestones. We strongly believe that they deserve a place here because of the symbiotic relationship they had with the — more narrowly defined — aspects of scientific development. Such a holistic approach towards the history of science, we feel, is not only justified, but essential.

- We started by saying that this book is intended for the lay-public and not for historians of science. But if you *do* want to know more about each of the topics covered here, we have provided ample references to the literature (with brief annotations) at the end of each chapter. Many of these works are quite erudite and scholarly and will certainly provide you with material to whet your appetite. But if you are happy with our story-telling, you can ignore the references! We have done the research for you and, in particular, whenever some of the incidences quoted in the text are controversial, we have taken care to alert you to this.

The development of science cannot be viewed in isolation, away from the social and political context in which it is taking place. You will find that we have not shied away from commenting on the influence of social, economic, and religious developments on science. Unfortunately, these influences have been rather negative in many crucial phases, but becoming aware of this fact is an important part of one's education. In a similar spirit, we have tried to portray scientists as normal human beings with normal human weaknesses and emotions (in spite of being the intellectual giants they were).

We have also tried to correct, whenever possible, the view that history is a frozen topic. This is not true about any branch of history and certainly not in the case of the history of science. A few of the historical facts described here — such as the development of calculus in south India, centuries before Newton and Leibnitz — have come to light only within the last few decades and hence are not as widely known as they deserve to be. To that extent, the history of science is very much alive and evolving in itself.

Many people have contributed at different stages to the making of this book. Several of the chapters overlap in their intellectual content with a series of articles one of us (TP) wrote for the journal *Resonance* in 2010–2012, even though they have all undergone a significant amount of rewriting and expansion. We thank the Indian Academy of Sciences for granting permission to Springer for the reuse of the material in these articles.

Many of our colleagues went through earlier drafts of this book and provided comments. We thank Jasjeet Bagla, Yashoda Chandorkar, V. Chelladurai, S.M. Chitre, Sunu Engineer, Harvinder Jassal, Kinjalk Lochan, Sunita Nair, J. V. Narlikar, Hamsa Padmanabhan, Krishnamohan Parattu, K. Srinivasan, and Tejinder Singh for their help.

Most of the figures have been reproduced from the public domain and we thank those who made them available in this manner for the wider use of the community. In addition, some figures in Chaps. 4, 5, 6, and 24 were reproduced with explicit permission from: Wellcome Library, London (Fig. 4.4), Collection of Michigan Medicine, University of Michigan Gift of Pfizer (Figs. 5.1, 5.2, and 5.3), Wolfgang Volk, Berlin (Fig. 6.1right) and the Special collections & Archives Research Center, Oregon State University Libraries (Fig. 24.4), and we are grateful to them for granting us this permission. (You will find the credit lines for all the figures, including those taken from the public domain, at the end of each chapter.) We also thank Manjiri Mahabal for the help she provided in her spare time as regards the organization of figures and figure credits.

Finally, it is a pleasure to thank Angela Lahee of Springer for all the assistance she has provided in this project. She encouraged us to bring out this book and was extremely helpful and supportive at every stage of its preparation. It has been a distinct pleasure to interact with her as a friend and work with her as a representative of Springer.

Pune, India
January 2019

Thanu Padmanabhan
Vasanthi Padmanabhan

Contents

1 In the Beginning ...

It is impossible to say when it was born. Maybe when the prehistoric hunter devised the perfect bow; or with the invention of the fire and the wheel. The true beginnings will probably never be known. Several simple acts of reasoning and innovation in the earliest moments are lost to us for lack of recorded history.

Even when some records exist, it is never easy to distinguish true science from mythology, magic, and mysticism. Every fable and fairy tale, myth and legend, epic and adventure story contains imaginative novelties. Did the creators of these works produce pure flights of fantasy or did they have first-hand experience with certain magnificent inventions? We believe it is the former and that the ancient literary fables cannot be taken seriously in determining the early development of science.

Given such a conservative stand, the earliest achievement of humanity requiring mammoth scientific and technological skill is definitely the pyramids in Egypt. These constructions, intended to ensure the safe passage of the Pharaohs (and some other privileged individuals) to the Other World, certainly needed elaborate technical skill. Such skill could not have been achieved without understanding the basics of what we today call 'science' [1].

These Egyptian pyramids, built around 2500 BC, were essentially funeral edifices. But their size, artistic structure, and technical perfection show the remarkable engineering skills of this ancient civilization. The earliest pyramid was probably the mastaba, a flat-topped rectangular structure with inward sloping sides made of mud-brick or stone with a shaft descending to the burial chamber far below it. The first mastaba, built by Pharaoh Djoser ($\sim$ 2670 BC), has a base measuring 120

T. Padmanabhan and V. Padmanabhan, *The Dawn of Science*,
https://doi.org/10.1007/978-3-030-17509-2_1

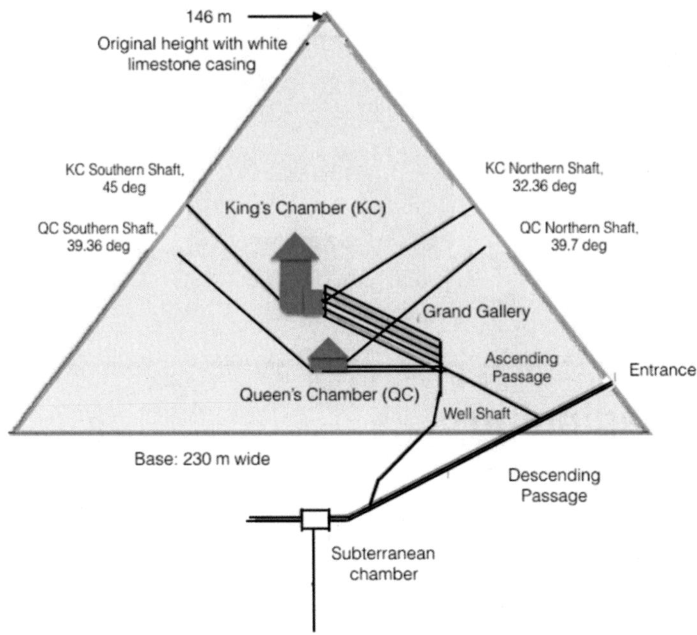

Fig. 1.1: The pyramid of Giza, one of the Egyptian engineering marvels. Based on the references to the fourth dynasty Egyptian Pharaoh Khufu in the interior chamber, Egyptologists estimate that the pyramid was built over a 10 to 20 year period around 2560 BC. Initially with a height of 146 metres, this was the tallest man-made structure in the world for more than 3800 years! As recently as 2017, scientists have discovered [2] a hidden chamber in the pyramid of Giza, the first such discovery of a new structure since the nineteenth century.

metres by 108 metres and a height of 60 metres. It started off as a mastaba tomb and evolved into a 60 metre high pyramid, having six layers, one built on top of the other, giving it a step pyramid structure.

Some of the earlier pyramids had a step-like structure. Usually, the builders started with a step pyramid, then added stones to pack and form a continuous slope covered with a smooth layer of limestone. The largest of these pyramids is the one at Giza, erected by the Pharaohs Khufu, Khafre, and Menkure. It has a square base of side 230 metres, a towering height of 146 metres, and comprises 2 300 000 blocks of stone; the King's chamber alone measures 10.5 metres by 5.2 metres in this pyramid (see Fig. 1.1).

Incidentally, while contemplating antiquity, we often tend to compress long periods of time in our minds and just think of them as 'ancient'. To avoid this trap, and maintain proper historical perspective, it is good to keep in mind the following fact: When the Greek historian Herodotus (484–425 BC) visited the pyramid of Giza, which he did, about 2500 years ago, he was actually acting like any tourist, looking at and admiring an 'ancient' tourist attraction. The pyramids were farther back in time for Herodotus than his visit is to us today!

It is rather difficult to identify *the* scientist who designed the pyramids; many people must have contributed. Nevertheless, an Egyptian scholar by the name of Imhotep, who lived around 2960 BC, has been named the architect of the 'step pyramid' at Saqqara in Egypt; this is probably the first major Egyptian structure in stone. In addition to being an excellent architect, he was also said to be a good vizier, mathematician, and medical man. There are also some ancient manuscripts, which suggest that Imhotep counseled Djoser (the Pharaoh of the step pyramid) about the seven-year famine in the Nile — a story that parallels the legends of Joseph in the Old Testament. Later, the Greeks identified Imhotep — called Imouthes in Greek — with Asklepios, the Greek god of medicine. This identification has made many historians doubt whether Imhotep really existed; probably Imhotep is no more real than Viswakarma — the celestial architect — appearing in the Indian epics. But the *man* who designed the step pyramid does deserve a mention when we describe the dawn of science.

Moving lightly forward over a thousand years, we come across the next milestone – the golden age of Egyptian mathematics and a scribe named Ahmose, who is mentioned in a mathematical treatise of ancient Egypt. Ahmose was a copyist who lived around 1000 BC and copied a mathematical treatise dealing with simple equations, fractions, and some of the rudimentary mathematical details from earlier, scattered works of anonymous origin. While some of the mathematical writings of the Egyptians show remarkable ingenuity (Box 1.1), they never generalized their methods and did not bother to develop mathematics as a scientific and logical discipline.

Box 1.1: Moscow Papyrus

The Moscow papyrus is actually an ancient Egyptian mathematical text, dating back to about 1850 BC, now kept in a museum in Moscow — other than that, it has no connection with Moscow! It contains a discussion of 25 problems occurring in different areas of mathematics. Of these, the discussion in problem 14 deserves special mention [3]. This attempts to calculate the volume of a 'frustum' of a square pyramid and height h, bounded between sections with *areas* x and y (see figure (A) below), not such an easy task.

Both the Babylonians and the Egyptians knew the answer to the corresponding problem in two dimensions, namely, what is the area of a truncated triangle with parallel *sides* a and b and height h (see figure (B) below). This is given by the formula $(1/2)h(a + b)$. Working by analogy, the Babylonians assumed the volume of the frustum of the pyramid to be $(1/2)h(x + y)$, where x, y are now the two areas, which turns out to be wrong. On the other hand, the Moscow papyrus shows that the Egyptians managed to 'guess' the correct formula: $(1/3)h(x + \sqrt{xy} + y)$! The intuitive leap involved in arriving at this formula is an important achievement of Egyptian mathematics [4].

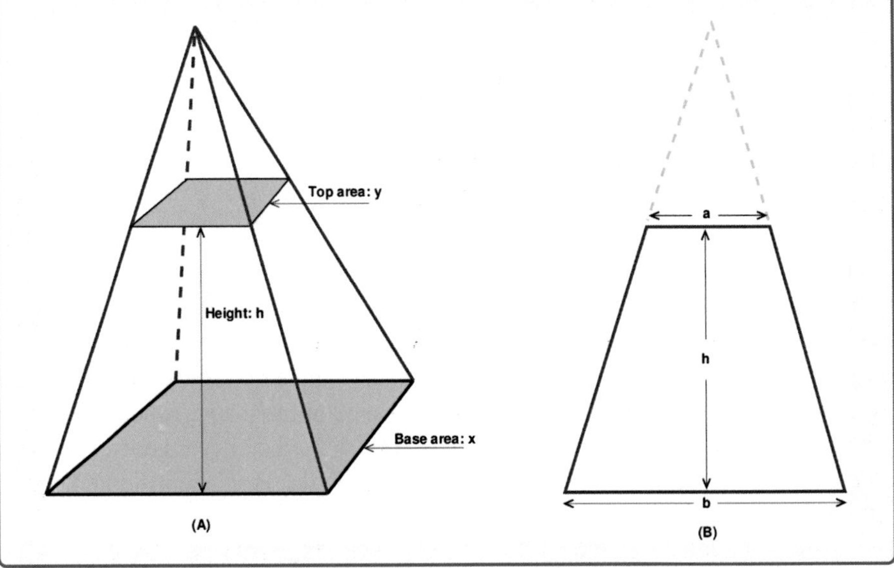

The development of mathematics as a scientific discipline had to wait for yet another thousand years — for Thales of Miletus (624–546 BC), who deserves to be called the founder of Greek science, mathematics, and philosophy [5]. Thales was born in Miletus, which is now located on the western coast of Turkey. He definitely travelled around, visiting Egypt and Babylonia, and learnt a lot from the Babylonians. Babylonians had, by then, developed the science of astronomy and had worked out detailed rules for calculating the eclipses and other astronomical phenomena. In fact, there is a story — coming to us from the historian Herodotus [6] — that Thales predicted a solar eclipse in 585 BC, thereby possibly preventing a war (see Box 1.2).

Box 1.2: A Solar Eclipse Could Stop a War!

Medes and the Lydians were engaged in a long-standing war and their armies were all set for yet another battle on 28 May 585 BC. Thales of Miletus, it is said, had predicted that a solar eclipse would occur on that very same day and, of course, it did indeed! The appearance of the eclipse was interpreted as an omen by both armies, and it stopped a battle between the Medes and the Lydians. Not only was the fighting immediately stopped, they even agreed to a truce.

This story comes to us from the ancient historian, Herodotus [6]. (How exactly Thales predicted the eclipse remains uncertain, and this has made some historians doubt the story.) If Herodotus's account is accurate, this eclipse is indeed the earliest non-trivial astronomical event recorded as being predicted in advance of its occurrence. Because astronomers can calculate the dates of historical eclipses, we know that the predicted eclipse was the solar eclipse of 28 May 585 BC. Isaac Asimov [7] called this battle the earliest historical event that is known *to the nearest day*, and described the prediction by Thales as "the birth of science"!

Thales also acquired concepts from Egyptian mathematics and developed it in a rather axiomatic way. He had the vision to understand the necessity of proof, and had developed a step-by-step logic leading to conclusions from given premises. This feat is truly incredible because — as far as we know — several other early civilizations like the Egyptians, Babylonians, Indians, and the Chinese seem not to have worried too much about this need for logical proof, which is the cornerstone of modern mathematics.

Thales, in fact, is credited with the following five elementary theorems: (1) The circle is bisected by any diameter. (2) The two base angles of an isosceles triangle are equal. (3) The vertically opposite angles made by two intersecting straight lines are equal. (4) Two triangles are congruent if they have two angles and the

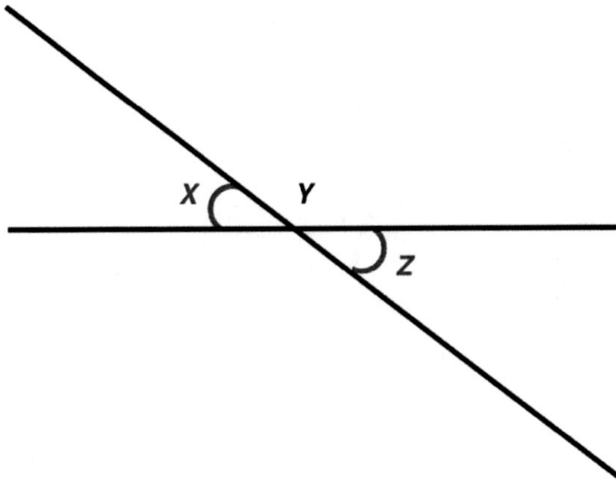

Fig. 1.2: Thales of Miletus (624–546 BC) deserves to be called the founder of Greek science, mathematics, and philosophy. Among other contributions, he was probably the first to understand the importance of the notion of logical proof for a theorem. While several other civilizations, like the Egyptians, Babylonians, Indians, and Chinese, had made progress in obtaining the correct mathematical results, they did not seem to have worried about the need for systematic proof for these assertions. The figure above gives an example of one of the five theorems Thales *proved*. The theorem illustrated by the figure states that the vertically opposite angles (x and z in the figure) made by two intersecting straight lines are equal. Thales provided the following elementary proof: "In the figure, x is the supplementary angle to y; but y is also a supplementary angle to z. Since things equal to the same thing are equal to one another, $x = z$."

corresponding side equal. (5) An angle inscribed in a semicircle is a right angle. These results were indeed known long before Thales, but were accepted as true on the basis of direct, empirical measurement. Thales' genius lies in proving each of these theorems by logical reasoning, starting from certain basic axioms. For example, he proved (3) in the following manner: "In Fig. 1.2, x is the supplementary angle to y; but y is also a supplementary angle to z. Since things equal to the same thing are equal to one another, $x = z$".

Thales also worried about more fundamental questions of nature like, "What is the Universe made of?" It is irrelevant that he didn't get the 'right' answer — we do not know the answer even today! What is important was that he asked the question and, for the first time in history, attempted to answer it without invoking gods, demons, and other mythological entities, which were available in plenty at the time. In this manner, he influenced his contemporaries as well as the coming generations in no small measure. So much so that when the Greeks listed their 'seven wise men', Thales was always ranked as the first.

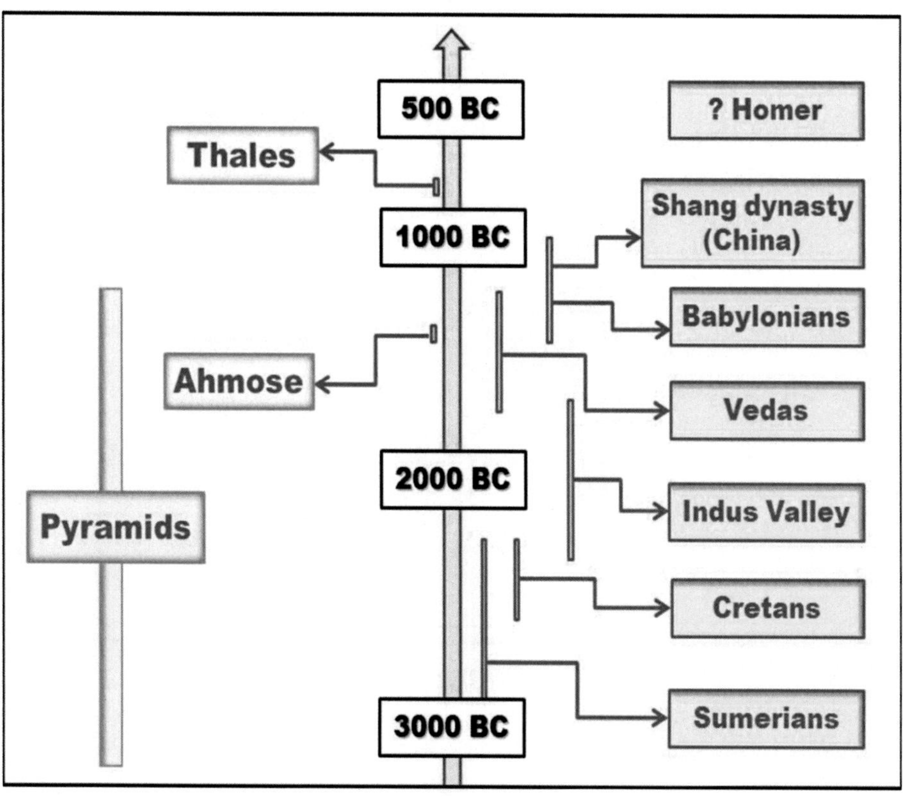

WHEN

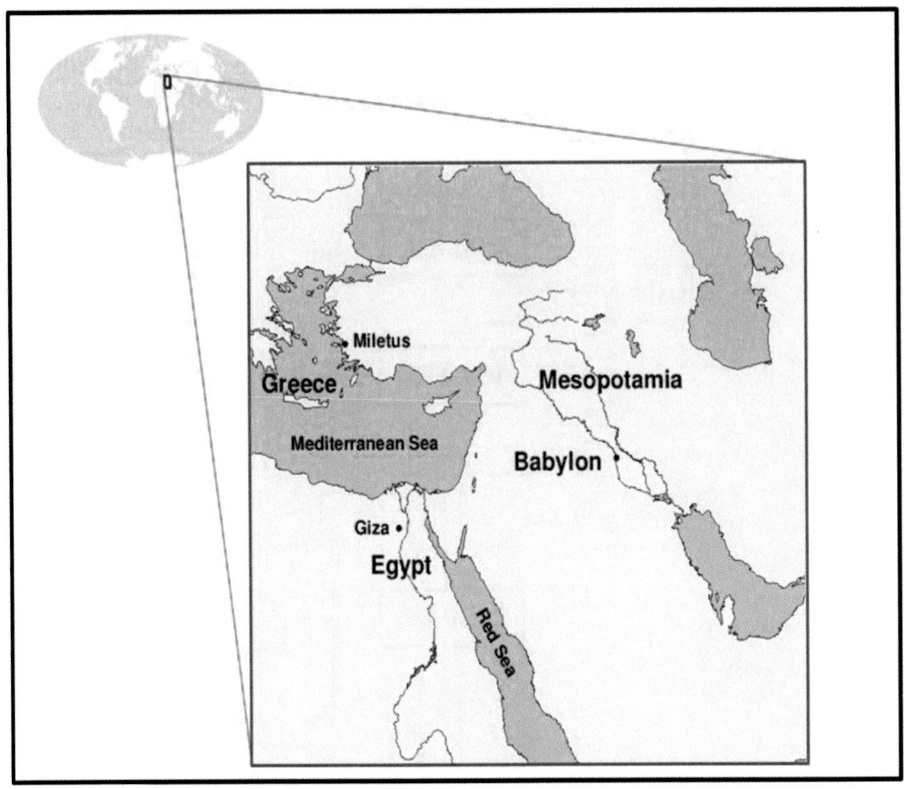

WHERE [8]

Notes, References, and Credits

Notes and References

1. For a comprehensive discussion of the pyramids in ancient Egypt, see:
Collins, Dana M. (2001), *The Oxford Encyclopedia of Ancient Egypt*, Oxford University Press, Oxford [ISBN 978-0-19-510234-5].
Jackson, K. and Stamp, J. (2002), *Pyramid: Beyond Imagination: Inside the Great Pyramid of Giza*, BBC Worldwide Ltd. [ISBN 978-0-563-48803-3].
Lehner, Mark (1997), *The Complete Pyramids*, Thames and Hudson, London [ISBN 0-500-05084-8].

2. We are still discovering various aspects of pyramids! See, for example, https://news.nationalgeographic.com/2017/11/great-pyramid-giza-void-discovered-khufu-archaeology-science/
3. Clagett, Marshall (1999), *Ancient Egyptian Science: A Source Book - Volume 3: Ancient Egyptian Mathematics*, Memoirs of the American Philosophical Society **232**, American Philosophical Society, Philadelphia [ISBN 0-87169-232-5].
4. The problems in the Moscow Papyrus are still being (re)interpreted by scholars; see, e.g., Cooper L. (2010), *A new interpretation of Problem 10 of the Moscow Mathematical Papyrus*, Historia Mathematica, **37**, 11–27.
5. For further reading, see, e.g.,
   Couprie, Dirk L. (2011), *Heaven and Earth in Ancient Greek Cosmology: from Thales to Heraclides Ponticus*, Springer, New York [ISBN 9781441981158].
   Luchte, James (2011), *Early Greek Thought: Before the Dawn*, Bloomsbury Publishing, London [ISBN 978-567353313].
   O'Grady, Patricia F. (2002), *Thales of Miletus: The Beginnings of Western Science and Philosophy*, Western Philosophy Series Vol **58**, Ashgate, New York [ISBN 9780754605331].
   Boyer, C.B. (1989), *A History of Mathematics*, Wiley, New York [ISBN 0-471-09763-2].
6. For an English version, see
   Grene, David (1987), *The History by Herodotus (translation)*, University of Chicago Press, Chicago, USA [ISBN 0-226-32770-1].
7. See, Asimov, Isaac (1991), *Isaac Asimov's guide to Earth and Space*, Fawcett Books, New York, p. 85 [ISBN 978-0-307-79227-3].

Figure Credits

8. The outline map of the world, used in the inset, in Chaps. 1, 3, 4, 6, 7, and 16 is courtesy FreeWorldMaps.net: http://www.freeworldmaps.net/outline/maps/contour-world-map.gif

2

The Athenian Contribution

Pythagoras ($\sim$ 570–495 BC), born on Samos, an Aegean island, emigrated to Croton in southern Italy around 529 BC. By that time, the coasts of southern Italy and eastern Sicily had already been colonized by the Greeks and had adopted the Greek way of life. Pythagoras founded a school in Croton, thereby extending the philosophical tradition of Thales — originally prevalent in eastern Greece — to the far west [1]. The members of the school debated mathematics, philosophy, and theology in great detail but — unfortunately — maintained a code of secrecy over their activities. This earned the school the disrepute of being a mystery cult, wedded to dangerous values, and brought it under active persecution, even during Pythagoras's lifetime. In fact, Pythagoras had to flee the city and lived his last ten years in voluntary exile.

This secrecy over their transactions also prevented later historians from ascertaining the Pythagorean contribution in an objective manner. None of the writings by Pythagoras himself has survived and his contributions have to be judged through references made by later thinkers. Based on such evidence, the following two contributions definitely deserve to be highlighted.

The first was the series of experiments Pythagoras conducted on the production of sound, by studying the notes emitted by plucking a stretched string. He quickly realized that sounds pleasant to the human ear are invariably associated with rational steps in the scale of notes. He also related the note of the sound emitted to the length of the plucked string. For example, if one string was twice as long as the

T. Padmanabhan and V. Padmanabhan, *The Dawn of Science*,
https://doi.org/10.1007/978-3-030-17509-2_2

other, the note it emitted was found to be one octave lower. These were probably the first ever set of systematic experiments conducted in any branch of physics.

These results led Pythagoreans to believe that the entire world could be constructed using simple, rational numbers. In fact, so enamored were they with this idea that they attributed several mystical properties to rational numbers! Though such ideas are too simplistic, they opened up the study of numbers — an active branch of modern mathematics. One can easily imagine their shock when they realized that there existed numbers which were not rational; that is, numbers which could not be expressed as the ratio of two natural numbers. If the side of a square is one unit, then the length of the diagonal represents an irrational number. Legend has it that Pythagoreans tried hard to keep the existence of such numbers a secret but the information leaked out — through some unfaithful members!

Box 2.1: Proving the Pythagoras Theorem

The 'standard' proof of this famous theorem is somewhat complicated. But it is possible to see the validity of this result without much ado if one uses some imagination. The figures below illustrate two such 'proofs without words' of this theorem (see also Box 3.2). The one on the left was known to the ancient Chinese; the one on the right was given by the Indian mathematician Bhaskara (1114–1185 AD) with the a simple comment 'Behold!' [2].

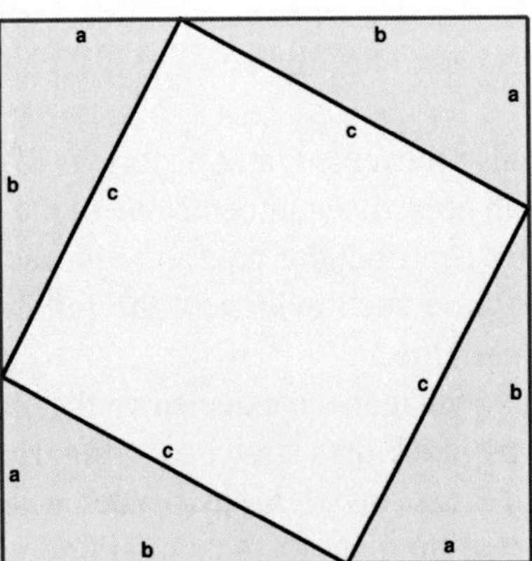

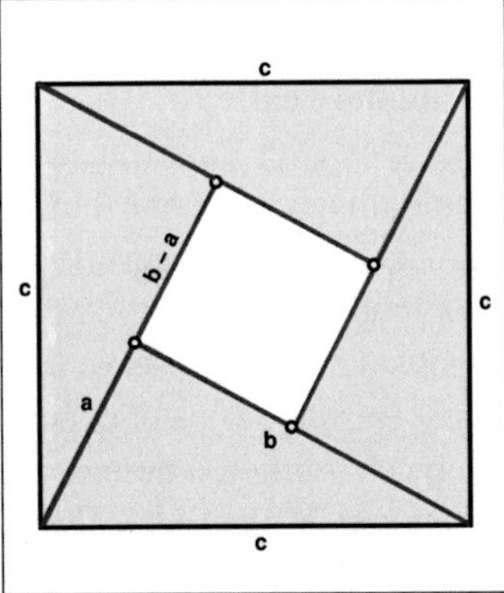

Pythagoras was also probably the first who taught that the Earth was spherical and that the orbits of the Sun, the Moon, and the planets were quite different from those of stars. He also guessed that the Morning Star (Phosphorus, as it was known) and the Evening Star (Hesperus) were in fact the same object. He named it Aphrodite, which we now call Venus. Pythagorean notions strongly influenced several later thinkers, like Anaxagoras (510–428 BC) — who taught at Athens for nearly 30 years and tried to put Pythagorean ideas on a more rational basis — and Democritus ($\sim$ 460–370 BC), who was the first to introduce the atomistic view of matter in the West. (See Box 16.2.)

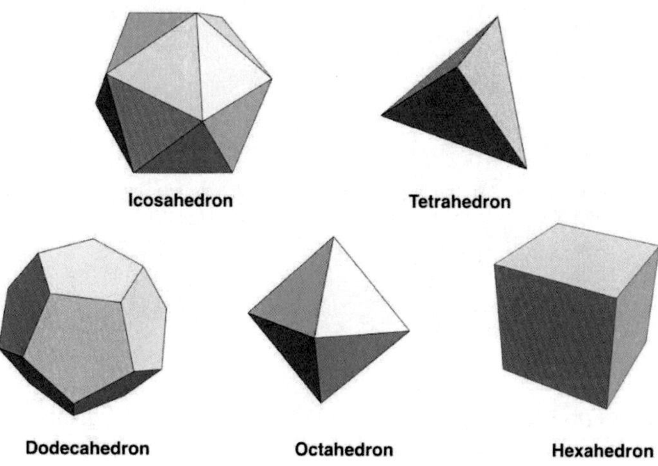

Fig. 2.1: Plato ($\sim$ 428–348 BC) knew that there were five — and only five — *regular* solids. A regular solid is defined to be one with identical faces, with all the sides and angles formed by the faces being equal. These five Platonic solids, shown in the figure, are the four-sided tetrahedron, the six-sided hexahedron (commonly known as the cube), the eight-sided octahedron, the twelve-sided dodecahedron, and the twenty-sided icosahedron. Plato, in fact, tried to model the entire natural world in terms these 'perfect' solids.

The most famous among those who hailed from Athens is, however, Plato (born around 428 BC and died around 348 BC) who was probably more a philosopher than a scientist [3]. (Incidentally, it is through him that we have come to know of most of the ideas of Socrates.) Plato did, however, have a fascination for mathematics as

Fig. 2.2: Plato ($\sim$ 428–348 BC) and Aristotle (384–322 BC) were the giants at the Academy in Athens and influenced views about Nature for centuries. The picture [7] is a portrayal of these two in the fresco "The School of Athens" by Raphael (painted between 1509 and 1511).

the "purest form of philosophy", and tried to use mathematical ideas to describe the heavens. He knew that there were five — and only five — regular solids. A regular solid is one with congruent faces, with all the lines and angles formed by the faces equal (see Fig. 2.1). These solids are the four-sided tetrahedron, the six-sided hexahedron (cube), the eight-sided octahedron, the twelve-sided dodecahedron, and the twenty-sided icosahedron. The Pythagoreans — and later on, Plato — believed that the five regular polyhedrons must play a crucial role in nature. They assumed that the 'atoms' of the four elements — fire, earth, air, and water — were in the shapes of the tetrahedron, cube, octahedron, and icosahedron, respectively. (In post-Aristotelian days, aether was introduced as the fifth element and was identified with the dodecahedron.) Plato tried to fit the entire natural world into a model based on these 'perfect' solids. This insistence, that the heavens should reflect human ideas of perfection, held sway in the ages to follow (see Box 2.2).

Plato established the famous Academy in Athens, which influenced the philosophical thinking of people all around the Mediterranean for years. It also remained the main stronghold of paganism in a Christian world in the following centuries, and was, in fact, ordered to be closed by Emperor Justinian in 529 AD, basically for this reason.

Box 2.2: Packing the Planets in Platonic boxes!

Today, we know that planetary systems around stars are ubiquitous and hundreds of stars in our galaxy have planets orbiting them. As such, the number of planets around a star and the radii of their orbits cease to have any special significance. But, in the medieval times, only five planets were known to mankind: Jupiter, Saturn, Mars, Venus, and Mercury.

Kepler (1571–1630 AD) discovered the laws governing their motion, and could relate the periods of their orbits around the Sun to their distances from the Sun. But he was intrigued as to what determines the distances of these planets from the Sun. Enamoured by the idea of Platonic solids, Kepler attempted to model the orbits of the known planets (Fig. 2.3) using the Platonic solids [4, 5].

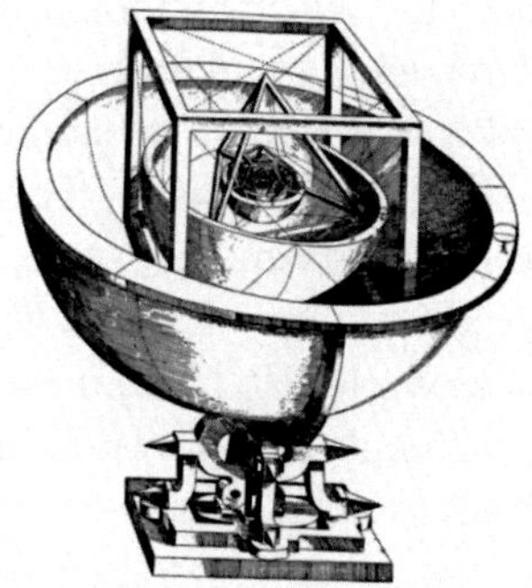

Fig. 2.3: Kepler's *Music of the Spheres* [8].

He started with a sphere with the same radius as the orbit of Saturn and inscribed a cube inside it. He then put another sphere inside the cube to represent Jupiter's orbit, put a tetrahedron inside this sphere with another sphere inside that to represent the orbit of Mars, a dodecahedron inside that to hold the terrestrial sphere, an icosahedron to hold the sphere of Venus, and, finally, an octahedron with a sphere inside that to represent Mercury's orbit. Interestingly enough, the radii of the orbits of the planets — calculated by this procedure — matched fairly well with the observed radii of the orbits! This is a classic example of a completely wrong model correctly explaining the observed facts!

The giant among the thinkers produced by the Academy was Aristotle (384–322 BC), who was considered the 'intelligence' of the Academy. As it turned out, Plato named someone else as his successor, and Aristotle quit the school in protest [6]! Aristotle was invited to Macedonia to tutor the King's son, Alexander (who later became Alexander the Great), which he did for about six years. He then returned to Athens and formed his own school, the Lyceum, where he lectured for nearly 12 years. Aristotle's lectures at this school constitute virtually a one-man encyclopedia of knowledge running over 50 volumes.

His best contribution was in the field of biology, in which he made a careful and meticulous classification of animal species, arranging over five hundred of them in different hierarchies. This classification scheme and related ideas were truly 'modern'. For example, he classified the dolphin with the beasts of the land because he knew dolphins nourished the fetus by a placenta. Later workers, who weren't as logical, put the dolphin back in the sea and it took nearly 2,000 years for biologists to set the record right! He also made careful observations of the developing embryo of the chicken and the complex structure of the stomach of the cow.

Aristotle's attempts in 'natural philosophy'— the branch of science we now call physics — however, were not so successful. For some strange reason, he did not use the experimental and observational approach — which was successful in biology — in the study of the physical world. He tried to explain the natural phenomena using the properties of five elements (earth, water, fire, air, and aether, the last of which was his own innovation), by attributing a 'natural place' to each element. He firmly believed, for example, that a heavier stone will fall to the ground faster than a lighter one, and — rather surprisingly — never bothered to check it. The fact that he was wrong has momentous consequences, and forms the basis for Einstein's general relativity!

Another recurrent theme we will meet in several later chapters, in which Aristotle's (wrong) views held sway for centuries, was in the description of the heavens and the motion of celestial bodies. For example, one such view concerned the cause for the daily rising and setting of the Sun (and the stars). The Pythagoreans had attributed this to the rotation of the Earth about its axis. And, in fact, one person in antiquity, Aristarchus of Samos ($\sim$ 310–230 BC), had even suggested a heliocentric model — in which the Earth, not only rotated on its axis, but also went around the Sun! However, Aristotle criticized the idea of *any* motion for the Earth; he advocated a heavenly sphere which contained fixed stars and rotated about the Earth. This was accepted by later astronomers — in particular, Claudius Ptolemy (100–170 AD) — who believed that the Earth would feel continuous gale force winds if it rotated.

Ironically enough, while Aristotle dominated later thinkers, he was not really as influential during his own time as, for example, Plato. His works were published only after his death, and — soon after the fall of Rome — were totally lost to Europe. These volumes of work, however, survived among the Arabs, who valued them dearly. Much later, in the twelfth and thirteenth centuries, Christian Europe rediscovered the Arabic texts and translated them into Latin. This led to Aristotle becoming the most influential amongst the ancient philosophers in medieval Europe.

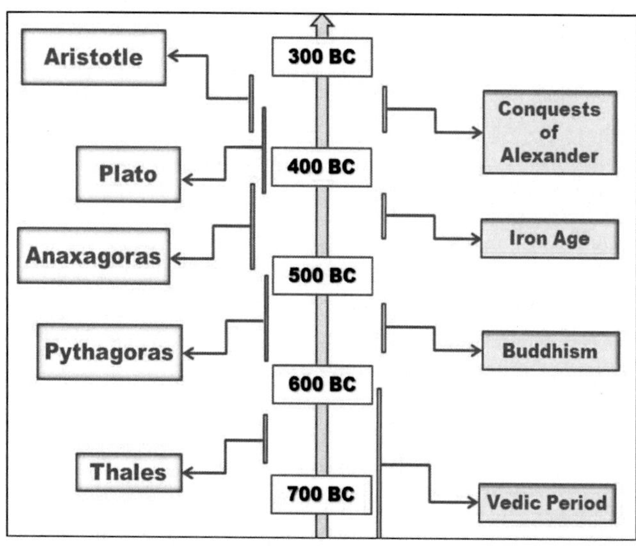

WHEN

Notes, References, and Credits

Notes and References

1. For further reading, see, e.g.,
   Riedweg, Christoph (2005), *Pythagoras: His Life, Teachings, and Influence*, Cornell University Press, New York [ISBN 978-0-8014-7452-1].
   Horky, Philip Sydney (2013), *Plato and Pythagoreanism*, Oxford University Press, Oxford [ISBN 978-0-19-989822-0].
   Joost-Gaugier, Christiane L. (2006), *Measuring Heaven: Pythagoras and his Influence on Thought and Art in Antiquity and the Middle Ages*, Cornell University Press, New York [ISBN 978-0-8014-7409-5].
   McKeown, J. C. (2013), *A Cabinet of Greek Curiosities: Strange Tales and Surprising Facts from the Cradle of Western Civilization*, Oxford University Press, Oxford [ISBN 978-0-19-998210-3].
   O'Meara, Dominic J. (1989), *Pythagoras Revived*, Oxford University Press, Oxford [ISBN0-19-823913-0].
2. For the origins of Pythagoras' theorem, see, e.g.,
   Ratner, Bruce (2009), *A Closer Look: Pythagoras: Everyone knows his famous theorem, but not who discovered it 1000 years before him*, Journal of Targeting, Measurement and Analysis for Marketing **17**, 229-242 [doi:10.1057/jt2009.16].
   Joseph, George Gheverghese (2010), *The Crest of the Peacock: Non-European Roots of Mathematics*, Third Edition, Princeton University Press, Princeton [ISBN9780691135267].
   Plofker, Kim (2009), *Mathematics in India*, Princeton University Press, Princeton [ISBN 9780691120676].
3. For more on Plato, see, e.g.,
   Jackson, Roy (2001), *Plato: A Beginner's Guide*, Hodder and Stroughton, London [ISBN 0-340-80385-1].
   Penner, Terry (1992), *Socrates and the Early Dialogues*, in 'The Cambridge Companion to Plato', Cambridge University Press, Cambridge, UK (pp. 121-169) [ISBN 9781139000574].
   Cooper, John M and Hutchinson, D.S. [Eds] (1997), *Plato: Complete Works*, Hackett Publishing Company, Inc., US [ISBN 0-87220-349-2].
4. There is a vast literature on the Platonic solids; see e.g.,
   Coxeter, H. S. M. (1973), *Regular Polytopes*, Dover Publications, New York [ISBN 0-486-61480-8].
   Gardner, Martin (1987), *The second Scientific American Book of Mathematical Puzzles and Diversions*, University of Chicago Press, Chicago; Chapter 1: The Five Platonic

Solids [ISBN 0226282538].
Pugh, Anthony (1976), *Polyhedra: A visual approach*, University of California Press, Berkeley [ISBN 0-520-03056-7].
Weyl, Hermann (1952), *Symmetry*, Princeton University Press, Princeton [ISBN 0-691-02374-3].
Lloyd, David Robert (2012), *How old are the Platonic Solids?*, Journal of the British Society for the History of Mathematics, **27**, 131–140 [doi:10.1080/17498430.2012.670845].

5. The application of Platonic solids to heavenly bodies by Kepler is described in:
Kepler, Johannes (1952), *The Harmony of the World* (translated by Charles Glenn Wallis), Encyclopedia Britannica.
Voelkel, J. R. (1995), *The music of the heavens: Kepler's harmonic astronomy*, Physics Today, **48**(6), 59–60.

6. While there is some controversy about this, most historians believe that Aristotle (like Xenocrates, another prominent member of the Academy), left because he was not chosen to succeed Plato as director of the Academy, with the position going instead to Plato's nephew, Speusippus. See, e.g.,
Anagnostopoulos, Georgios (Editor) (2009), *A Companion to Aristotle*, Chap. 1, p. 6; Wiley, Sussex, UK [ISBN 978-1405122238].
Lloyd, G. E. R. (1968), *Aristotle: The Growth and Structure of his Thought*, Cambridge University Press, Cambridge, UK [ISBN 9780521094566].

Figure Credits

7. Figure 2.2 courtesy: Raphael / Wikimedia Commons / https://commons.wikimedia.org/wiki/File:Sanzio_01_Plato_Aristotle.jpg (from public domain).

8. Figure 2.3 courtesy: Johannes Kepler/ Wikimedia Commons/ https://commons.wikimedia.org/wiki/File:Kepler-solar-system-1.png (from public domain).

3

From the Ishango Bone to Euclid

Even a prehistoric tribe would have needed the notion of counting things — for example, to make sure that all the cattle came back home, or that the home tribe outnumbered its enemies. The most primitive (and, rather ironically, the most sophisticated) form of counting involves setting up a one-to-one correspondence between the objects that are to be counted and some other convenient set of objects like, for example, the fingers on one's hand. Even today, primitive African hunters keep count of the number of wild boar they kill by collecting the tusks of each animal, and young girls in the Masai tribe — who live on the slopes of Mt. Kilimanjaro — wear brass rings around their necks, one for each year of their age [1].

This process of counting soon evolved into a more detailed process of keeping 'tally marks' on a bone or a stone. One such bone, called the 'Ishango bone', belonging to the period from 9000 to 6500 BC, was found at the fishing village of Ishango (on the shores of Lake Edward in Congo) in 1962. If the markings on this bone are indeed tally marks — as many historians believe — this is probably the earliest available record of mathematical activity [2].

From such humble beginnings, mathematics progressed fairly rapidly. Counting made primitive men realize that 'one cow and one cow make two cows' just as 'one spear and one spear make two spears'. To extract, from such a concrete experience, the abstract idea that 'one and one make two' was the *first major breakthrough in abstract mathematical thought.*

T. Padmanabhan and V. Padmanabhan, *The Dawn of Science*,
https://doi.org/10.1007/978-3-030-17509-2_3

This idea, which appears so obvious to us today, involves thinking of 'one' and 'two' as independent abstract entities existing on their own without being associated with cows or spears. That is, we stop asking "two what?" when someone says "two"! In fact, this abstraction has not yet been achieved by some tribal societies even today. For example, Fiji tribals distinguish ten boats (which they call bole) from ten coconuts (koro) and have a separate name, saloro, for one thousand coconuts!

In practical terms, this abstraction requires the words in a language to describe these numbers as separate nouns. All ancient civilizations — in particular, the Egyptian, Chinese, Babylonian and Indian — developed such verbal descriptions for numbers at some stage in their development. In the early days, the words for numbers often originated from parts of the body, fingers, heads, etc., and could describe only small numbers. In fact, words originally existed only for 1 and 2, somewhat rarely for 3, and then 'many'. For example, Egyptian and Chinese writings sometimes even identified 3 with many! The Egyptian word for water was just the word for wave repeated three times; in Chinese, the 'forest' is identified with '3 trees', and 'fur' with '3 hairs' and — very chauvinistically — 'crowd' is '3 men'!

The written forms of these 'number words' obviously constituted the earliest mathematical notation. Very soon, they were condensed into more compact and useful forms. The most primitive, and fairly complicated, notation among them which has — surprisingly enough — survived till this day are the Roman numerals: I, V, X, L, C, etc. The most useful, of course, are the Indo-Arabic numerals, which we will describe in a later chapter (see Chap.7). Other civilizations had their own symbols.

A crucial feature of the way numbers are represented is the development of the positional notation and the concept of zero. The positional notation allows one to assign a different numerical value to the same symbol depending on where it occurs in a string of numbers. For example, we think of the number 8 occurring in 89 to have a different value than the 8 occurring in 28. This idea is very ancient and virtually every well-developed ancient civilization had developed some form of positional notation.

The next important step is the concept of zero which actually has *two* subtly different aspects. The first is the need for a symbol for zero to distinguish 304 from 34 in the positional notation. Different civilizations handled this in different ways, using, for example, an empty space, a dot, or a circle as a symbol for zero. The second, and somewhat more advanced, aspect is to recognize zero as a number in its own right, like say 3 or 5, and have definite rules for manipulating it. It appears that this aspect of zero — viz., the fact that it is a number rather than just a placeholder in the positional notation — was recognised in Indian mathematics fairly early on. An early work which discusses zero as a number is the *Brahma-sphuta-siddhanta* of the Indian astronomer and mathematician Brahmagupta (598–668 AD), written around 628 AD (see Box 3.1).

Box 3.1: Dealing with Nothing!

Expressing arbitrarily large numbers using a small set of symbols requires tremendous ingenuity and involves three distinct ideas: (i) The positional notation in which the value of a symbol is not fixed but depends on its location; for example, we interpret 23 as '2 tens' and '3 ones' and 32 as '3 tens' and '2 ones'. (ii) A convenient choice for the 'base' to use in the positional notation; we normally use 10 as the base, so that 467 will stand for $4 \times 100 + 6 \times 10 + 7 \times 1$. And most importantly, (iii) a symbol for 'nothing' (0) which allows us to distinguish 203 from 23.

Different ancient civilizations achieved varying degrees of success in this task. The Babylonians, the Egyptians, the Chinese, and the Indians all had the idea of a positional notation. As early as 3000–2000 BC, the Babylonians developed a positional system with a base of 60 (reflected even today in the division of hours, minutes, and seconds)! However, to avoid having to use separate symbols for 1 and 59, they also used a grouping idea based on 10. The Indian Kharosti numerals, on the other hand, involved the use of three different groupings of 4, 10, and 20. The Chinese, whose language provides a separate pictographic character for each idea, used a notation which explicitly spelt out the value of the number.

The last crucial step is to have a symbol for nothing. There is an ancient Indian manuscript, called the *Bakhshali* manuscript, which was a practical manual on arithmetic for merchants, that contains the symbol for zero. This manuscript is currently at the University of Oxford's Bodleian Libraries. The carbon dating of three pieces of this manuscript [3] has, unfortunately, led to widely disparate dates (224–383 AD, 680–779 AD, and 885–993 AD) but some historians believe, based on other internal evidence, that the earlier dates are more likely to be correct. There is also an occurrence of the symbol for zero in an inscription found in Cambodia [4] which is considered to have been made around 680 AD. Several inscriptions from this region are based on Sanskrit, and the Indian calendar year (called the Saka era; see Box 21.4) is often mentioned in these inscriptions. This indicates the possibility that ancient traders and settlers in this region who came from India brought with them the Sanskrit language, the Indian Saka era, the Sanskrit words for numbers, etc.; if so, they may also have brought along the concept of zero.

While all these examples are clouded in uncertainty, there is a *definitive* occurrence of zero in a text inscribed on the wall of a small temple near Gwalior, India, going back to AD 870. The temple inscription lists, among the various gifts given by the king to the temple, "a tract of land 270 royal hastas long and 187 wide, for a flower garden". This tells us that zero was in common usage in India prior to this date.

The *sunya*, invented in India, reached the West through the Arabs (Fig. 3.1), becoming zero in the process [5].

Next to counting, the ancients were preoccupied with sizes and shapes, which naturally led to the development of the science we now call geometry. The Greeks, especially Thales and Pythagoras (described in Chaps. 1 and 2), contributed significantly to the development of geometry. But there is one name which stands out above the rest, and whose work has exerted a lasting influence over the centuries — Euclid ($\sim$ 300 BC) of Alexandria.

After the death of Alexander in 323 BC, the Macedonian Empire was divided into three, and the Egyptian region came under the rule of Ptolemy Soter, whose dynasty continued for nearly 250 years, ending with Cleopatra. He chose Alexandria as his capital and opened the gates of the University of Alexandria to scholars from all over, making this city a hub of academic activity for centuries. One of the scholars in Alexandria was Euclid, the mathematician.

Very little is known about his life, except that he lived around 300 BC, taught for a few years at the university, and, most importantly, compiled a monumental 13-part treatise called *The Elements*. This work, which has dominated the teaching of (and thinking about) geometry for the past two thousand years, definitely earns him a special mention in the annals of science [6].

We do not possess a copy of *The Elements* from Euclid's own time. All modern editions are based either on a version produced by Theon of Alexandria (a Greek commentator who lived 700 years after Euclid) or from an anonymous compilation found in the Vatican library. Greek commentaries on Euclid were translated by three Arabic scholars at different times during the Middle Ages (Fig. 3.2). From the Arabic, it was translated to Latin; the first Latin translation was made in AD 1120 by Adelard of Bath — who actually had to travel to Spain disguised as a Muslim student, to obtain an Arabic copy! In all these versions, *The Elements* comprises 13 books with a total of 465 theorems.

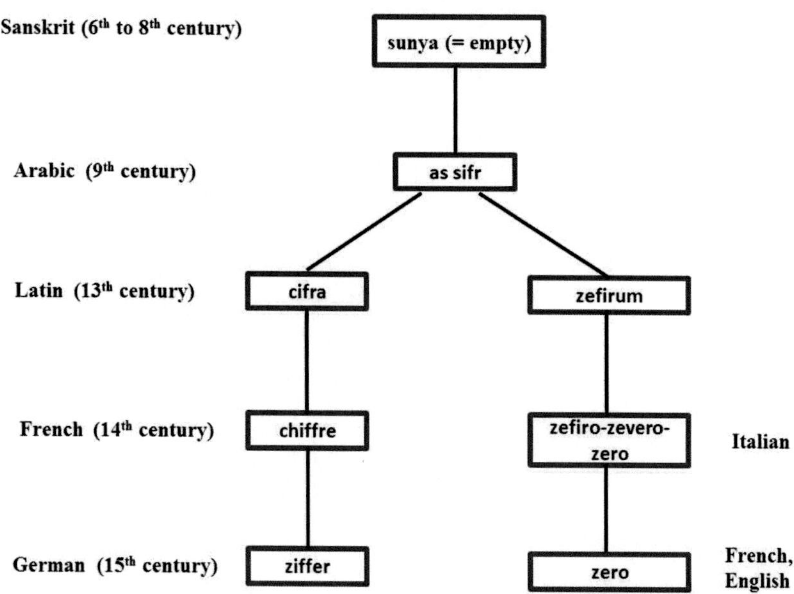

Fig. 3.1: Many ancient civilizations, including the Babylonians, the Egyptians, the Chinese, and the Indians had the idea of a positional notation, in which the value of a number symbol depends on its position in a string, e.g., the symbol 4 carries different values in 45 and 54. The next important step was to distinguish 405 from 45, which is possible only if there is a symbol for 'nothing'. This development seems to have come much later, with various choices for representing 'nothing', like an empty space, a dot, etc., being used at different periods of time. The earliest, definitive occurrence of zero is in a text engraved on the wall of a small temple near Gwalior (in India) in AD 870. This inscription lists various gifts given by the local king to the temple, including "a tract of land 270 royal hastas long and 187 wide". This symbol, called *sunya*, was developed in India, and it reached the West through the Arabs, becoming zero in the process. The figure shows this transmutation of (Indian) sunya to (modern) zero

Book 1 begins with the basic axioms and develops the theorems about the congruence of triangles, parallel lines, and rectilinear figures. (Theorem 47, for example, is Pythagoras' theorem, discussed in Chap. 2.) Book 2 deals with the algebraic results arising from the Pythagorean constructions, while Book 3 deals with several results for circles, tangents, and secants. Books 4, 5, and 6 discuss

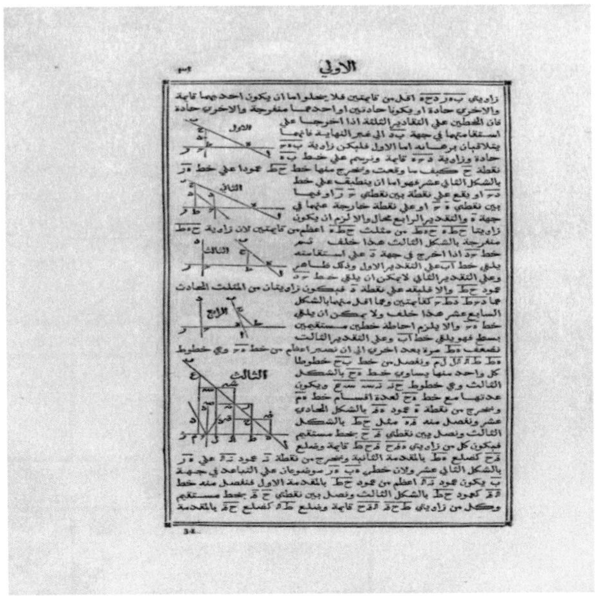

Fig. 3.2: Euclid of Alexandria, who lived around 300 BC, compiled a monumental 13-part treatise on geometry, called *The Elements*. This work has dominated the teaching of geometry for the past two thousand years. The original manuscripts from the days of Euclid are not available but, nearly 700 years after Euclid, the Greek scholar, Theon of Alexandria, wrote a commentary on *The Elements*. This is one of the two sources available to us; the other being an anonymous compilation found in the Vatican library. The Greek commentaries were translated into Arabic by three Arabic scholars at different times during the Middle Ages. The above figure shows an Arabic version (made around AD 1258) of Euclid's Elements [9]. The Arabic works were translated into Latin in AD 1120 by Adelard of Bath. This is just one of the many examples of the Greek contributions being preserved and then reappearing in Europe in later centuries through the legacy of the Arab civilisation.

various geometrical constructions and similarity of figures, and the last three books (11, 12, and 13) contain theorems on solid geometry. Probably the most remarkable — and also the least known — volumes are Books 7 to 10. These discuss, not pure geometry, but elementary results in number theory! In fact, they contain some of the most fundamental results in this subject.

This compilation, of course, was based on the earlier contributions of several people. Before Euclid, a previous compilation by Theudius was used at Alexandria. Euclid enlarged this work, drawing heavily on material already developed by Theudius, Exodus, and Hippocrates of Cos — all of whom contributed to different aspects of geometry and number theory. In fact, some cynical historians have commented that Euclid's exposition was brilliant only in those areas in which he had excellent sources at his disposal. But even as a compilation, *The Elements* is a notable achievement. The pattern of logic and order in this treatise was completely due to Euclid, and this in itself was an outstanding contribution to the evolution of the subject. Euclid might not have been a first-rate mathematician but he was a first-rate teacher, and his book remained in use, practically unchanged, for nearly 2000 years. Euclid is probably the most successful textbook writer of all times, with *The Elements* having gone through more than 1000 editions after the invention of printing!

Box 3.2: Two Ancient Gems from the Orient

There are two other ancient mathematical texts, both of Oriental origin, which contain several interesting results. These are the Indian treatise called *Sulvasutras* [7], dated by various historians to anywhere from 800 BC to 100 BC, and the Chinese work *Chiu Chang Suan Shu* (Nine Chapters on the Mathematical Art), probably written around 250 BC.

The *Sulvasutras* contains, among other things: (1) an explicit statement of what we call the Pythagoras theorem (in terms of the length, breadth, and diagonal of a rectangle); (2) several examples of Pythagorean triples — which are integers (a, b, c) satisfying the relation $a^2 + b^2 = c^2$, and (3) the construction for producing a square equal in area to a given rectangle (a problem, incidentally, which arises in making a falcon shaped altar for sacrifices!). What is probably equally important is a discussion, found in the text Apastamba Sulvasutra, of the area of a trapezium with bases 24 and 30 units and width 36 units. The ancient text not only calculates the area correctly, but also gives a purely geometrical (Euclid-style) proof for the result!

The Chinese work [8], 'Nine Chapters', deals with a gamut of problems in simple mathematics: operations with fractions, measurement of areas of rectangles, trapezia, triangles (which are done correctly), circles, segments and sectors of circles (which are done, approximately with π taken as 3!), volumes of elementary solids (including the frustum of the pyramid), extraction of square and cube roots, and a system of linear equations.

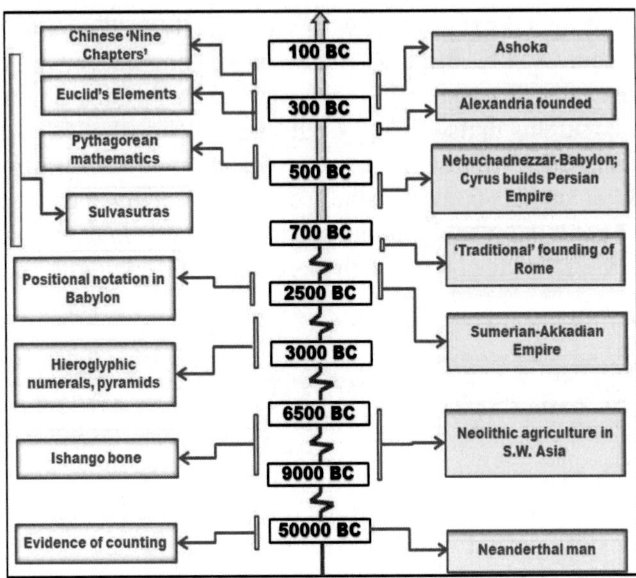
Chinese 'Nine Chapters'
Euclid's Elements
Pythagorean mathematics
Sulvasutras
Positional notation in Babylon
Hieroglyphic numerals, pyramids
Ishango bone
Evidence of counting
100 BC
300 BC
500 BC
700 BC
2500 BC
3000 BC
6500 BC
9000 BC
50000 BC
Ashoka
Alexandria founded
Nebuchadnezzar-Babylon; Cyrus builds Persian Empire
'Traditional' founding of Rome
Sumerian-Akkadian Empire
Neolithic agriculture in S.W. Asia
Neanderthal man

WHEN

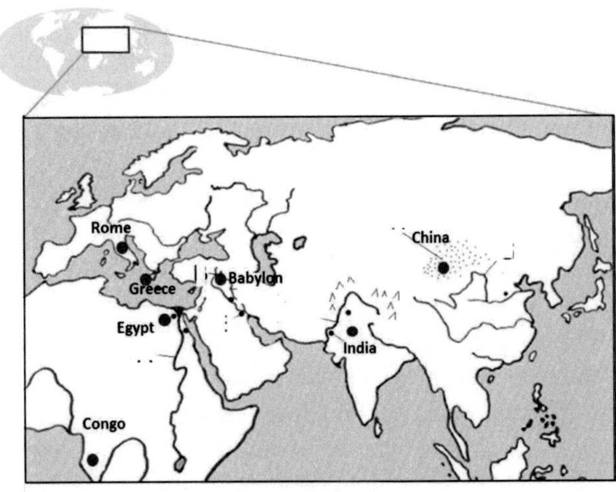
Rome
Greece
Babylon
Egypt
India
China
Congo

WHERE

Notes, References, and Credits

Notes and References

1. This is claimed in page 53 of the book:
   Menninger, Karl (1969), *Number Words and Number Symbols*, Dover Publications, New York [ISBN 978-0262130400],
   but has been criticized in the work:
   Zaslavsky, Claudia (1999), *Africa Counts: Number and Pattern in African Cultures*, Chicago Review Press, Chicago, USA [ISBN 978-1556523502].
   See also:
   Hollis, A. C. (1905), *The Masai: their language and folklore*, Clarendon Press, Oxford, UK [ISBN 052404340X (microfiche)]
   which suggests that the custom, while prevalent, is not universally linked to the age of the woman.
2. The Ishango bone is described in many books on the history of mathematics. See, e.g.,
   Rudman, Peter Strom (2007), *How Mathematics Happened: The First 50 000 Years*, Prometheus Books, US [ISBN 978-1-59102-477-4].
   de Heinzelin, Jean (1962), *Ishango*, Scientific American, **206** (June 1962) 105–116.
   Stewart, I., Huylebrouck, D., Horowitz, D. et al. (1996), *The Bone that Began the Space Odyssey*, The Mathematical Intelligencer, **18**, 56
   [https://doi.org/10.1007/BF03026755].
3. See the document: *Carbon dating finds Bakhshali manuscript contains oldest recorded origins of the symbol 'zero'*; Bodleian Library (2017-09-14) at the URL http://www.bodleian.ox.ac.uk/bodley/news/2017/sep-14
4. See e.g.,
   Aczel, Amir (2015), *Finding Zero: A Mathematician's Odyssey to Uncover the Origins of Numbers*, Palgrave Macmillan, New York [ISBN 978-1137279842].
   Joseph, George Gheverghese (2016), *Mathematics: Engaging with the World from Ancient to Modern Times*, World Scientific, Singapore [ISBN 9781786340603].
5. For more on the history of zero and positional notation and related topics, see, e.g.,
   Kaplan, Robert (2000), *The Nothing That Is: A Natural History of Zero*, Oxford University Press, Oxford[ISBN 978-0195142372].
   Berggren, J. Lennart (2007), *Mathematics in Medieval Islam in The Mathematics of Egypt, Mesopotamia, China, India, and Islam: A Sourcebook*, Princeton University Press, Princeton [ISBN 978-0-691-11485-9].
   Georges, Ifrah (1988), *From One to Zero: A Universal History of Numbers*, Penguin Books US [ISBN 0-14-009919-0].
   Graham, Flegg (2002), *Numbers: their history and meaning*, Courier Dover Publications, US [ISBN 978-0-486-42165-0].

Dantzig, Tobias (1954), *Number–The Language of Science* (4th edn.), The Free Press (Macmillan Publishing Co.), UK [ISBN 0-02-906990-4].

6. For more on Euclid, see:
   Artmann, B. (1999), *Euclid: The Creation of Mathematics*, Springer, New York [ISBN 978-0-387-98423-0].
   Heath, T. (1956), *Euclid: The Thirteen Books of The Elements*, Dover Publications, New York [ISBN 978-0486600888].
   Mlodinow, L. (2002), *Euclid's Window: The Story of Geometry from Parallel Lines to Hyperspace*, Touchstone, New York [ISBN 978-0684865249].
   Knorr, Wilbur Richard (1975), *The Evolution of the Euclidean Elements: A Study of the Theory of Incommensurable Magnitudes and Its Significance for Early Greek Geometry*, Dordrecht, Holland [ISBN 90-277-0509-7].
7. More on the ancient Indian contribution can be found, e.g., in:
   Divakaran, P. P. (2018), *The Mathematics of India: Concepts, Methods, Connections*, Springer Verlag, Singapore [ISBN 978-9811317736].
   Balachandra Rao, S. (1998), *Indian Mathematics and Astronomy: Some Landmarks*, Jnana Deep Publications, Bangalore, India [ISBN 81-900962-0-6].
   Joseph, George Gheverghese (2011), *The Crest of the Peacock: Non-European Roots of Mathematics*, Third Edition, Princeton University Press, Princeton [ISBN 9780691135267].
   Plofker, Kim (2007), *Mathematics in India in The Mathematics of Egypt, Mesopotamia, China, India, and Islam: A Sourcebook*, Princeton University Press, Princeton [ISBN 978-0-691-11485-9].
   Sarma, K.V. (1997), *'Sulbasutras'*, in *Encyclopedia of the History of Science, Technology, and Medicine in Non-Western Cultures*, edited by Helaine Selin, Springer, New York [ISBN 978-0-7923-4066-9].
8. For more details on the Chinese contribution, see:
   Dauben, Joseph W. (2007), 'Chinese Mathematics', in Victor J. Katz, *The Mathematics of Egypt, Mesopotamia, China, India, and Islam: A Sourcebook*, Princeton University Press, Princeton [ISBN 978-0-691-11485-9].
   Martzloff, Jean-Claude (1996), *A History of Chinese Mathematics*, Springer, [ISBN 3-540-33782-2].
   Needham, Joseph (1959), *Science and Civilization in China: Volume 3, Mathematics and the Sciences of the Heavens and the Earth*, Cambridge University Press, Cambridge [ISBN 978-0521058018].

Figure Credits

9. Figure 3.2 courtesy: Euclid and Nasir al-Din al-Tusi / wikimedia commons/ https://commons.wikimedia.org/wiki/File:Houghton_56-1235_-_Euclid,_Elements_%28Arabic%29,_1594.jpg (from Public domain); [Source: "*56-1235, Houghton Library, Harvard University"].

4

Archimedes — the Giant Among the Ancients

A tombstone was discovered in the courtyard of the Hotel Panorama in Syracuse in the early 1960s while doing some excavation work. This was claimed to be the tombstone of Archimedes, who wanted his tombstone to bear the picture shown in the inset in Fig. 4.1. (See the top left part of the figure.) Archimedes (∼ 287–212 BC) was one of the greatest scientists who ever lived. Of this calibre there have been only three more since then: Galileo, Newton, and Einstein [1].

It is rather amusing that such a scientist did not come from the centre of intellectual activity at the time — Alexandria (see Chap. 2). He was born in Syracuse, around 287 BC. His father was a well-known astronomer of considerable talent. Archimedes did spend some time in Alexandria training under Euclid's students, but he soon returned to his native town, possibly because of his close friendship with the King of Syracuse, Hieron II.

Archimedes thrived intellectually in Syracuse. No other scientist of ancient times, not even Thales, inspired so many legends; the most famous, of course, is the one about 'Eureka'. Archimedes was responsible for several inventions and discoveries in various branches of science, including what we now call mechanics, hydrodynamics, optics, and pure mathematics. The details of his work are spelt out in nine Greek treatises which, fortunately, are still available to us.

Several others before him had conjectured about levers, but Archimedes worked out the principle of the lever in rigorous mathematical terms: "A small weight at a greater distance from a fulcrum can balance a large weight nearer the fulcrum." This led to the science of statics and to the useful notion of the center of gravity of bodies.

T. Padmanabhan and V. Padmanabhan, *The Dawn of Science*,
https://doi.org/10.1007/978-3-030-17509-2_4

Fig. 4.1: Legend has it that Archimedes wanted his tombstone to have the picture (given in the inset) of a sphere inscribed in a cylinder. It symbolises his brilliant approach — as described in the treatises *On Sphere and Cylinder* and *Method* — to show that the volume of a sphere is $(4\pi R^3/3)$ and its surface area is $4\pi R^2$. To obtain these results, he had to think of a finite sized solid as being made of an infinite number of infinitesimally small pieces. This idea of manipulating *infinitesimal* quantities lies at the heart of calculus (Box 20.1) and Archimedes almost invented this method! In 75 BC, about 137 years after the death of Archimedes, the Roman orator Cicero was serving as quaestor in Sicily and wanted to locate the tomb of Archimedes. It is claimed that Cicero eventually found it near the Agrigentine gate in Syracuse, in a dilapidated condition and got it resurrected [2]. The figure shows an artist's depiction of this event [7]. The painting was done in 1805, by Benjamin West (1738–1820), an Anglo-American history painter.

Two of the volumes authored by Archimedes — *On the Equilibrium of Planes* and *On Floating Bodies* — elaborate on the implications of these concepts. In these volumes, Archimedes spends considerable time establishing the positions of equilibrium of floating bodies of various shapes; the results are of great importance in naval architecture.

Archimedes did use these principles in several practical devices. He is supposed to have perfected a hollow cylinder of helical shape which, when rotated, could

Fig. 4.2: Archimedes is supposed to have realised that an object, if placed in a fluid, displaces a weight of fluid equal to its own weight, when he climbed into a bath tub, leading to the famous 'Eureka' incident [8]. The story goes that he put this idea to good use in determining the purity of the gold crown of the King of Syracuse. This incident is described by Vitruvius, a Roman writer, who included this tale in the introduction to his book on architecture, around the first century BC, nearly 200 years after the event is presumed to have taken place. (Many historians doubt whether the incident actually took place.)

serve as a water pump. He also devised a heavenly globe and a primitive model of a planetarium which could depict the motion of the planets. In his heart of hearts, however, he was a purist and did not really care too much for these applications. The contributions he was most pleased with were the results he could obtain in the branch of pure mathematics — in the determination of areas and volumes of geometrical shapes. He came up with very ingenious arguments — described in the treatises *On Sphere and Cylinder* and *Method* — to show that the volume of a sphere of radius R is $(4\pi R^3/3)$ and its surface area is $4\pi R^2$. To obtain these results he had to use the notion of a solid being made up of a large number of extremely small pieces and take — what we now call — limits of functions. If only he had used more compact and consistent notations, he might have discovered integral calculus! Another contribution was in the development of a systematic technique for computing the numerical value of π — which was used by several later workers as well (see Box 4.1).

Box 4.1: A Piece of Pi

Every ancient civilization which had to build anything of significance needed to know the length of the perimeter of a circle of a given diameter. They all knew that the ratio between the circumference and the diameter was a constant; and was roughly equal to 3. The task was to determine it exactly [3].

Some civilizations (like the ancient Hebrews) were quite happy with a value of 3, while others wanted something more accurate; the Egyptians, for instance, used $\pi = 22/7$ and the Chinese had the value $\pi = 355/113$. It was Archimedes who devised a *systematic* method that allowed one to compute the value of π to *any* desired accuracy (see also Box 20.2).

His idea was to take the circle and inscribe and circumscribe polygons, then measure the perimeters of these polygons, a fairly straightforward procedure. As the number of sides of the polygons increases, the circle remains wedged between them (see figure below) and the perimeters of the polygons provide a good approximation to the perimeter of the circle. With extraordinary patience, Archimedes used polygons with 96 sides and obtained the value for π to be $(3123/994) = 3.14185\ldots$, which is off the correct value by only 1 part in 12 500!

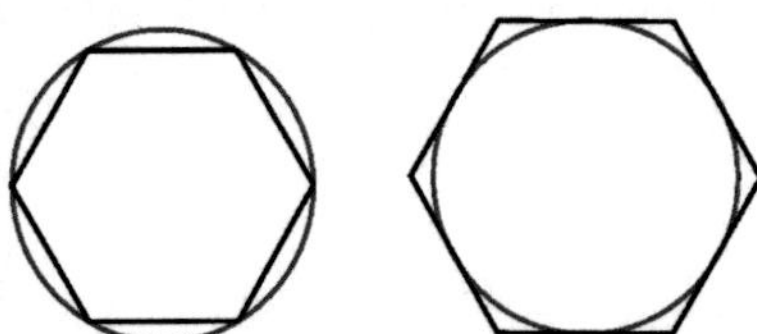

Unfortunately, Archimedes could not end his life in peace. During the time when Hieronymus was the ruler of Syracuse, the Romans (with sufficient provocation, it would seem; see Box 4.2) sent a fleet commanded by Marcellus to lay siege to Syracuse. This is supposed to have started the strange three-year war between the mighty Roman fleet on one side and a one-man-army, viz., Archimedes, on the other side. The city, however, fell after three years and Archimedes was killed

(212 BC) by a Roman soldier (Fig. 4.3) — apparently much to the disappointment of Marcellus, who did not want him to be killed. It is said that Marcellus honoured Archimedes with a proper funeral and a tombstone as Archimedes had requested [4].

In 75 BC, about 137 years after the death of Archimedes, the Roman orator Cicero was serving as quaestor in Sicily. Having heard stories about the tomb of Archimedes, he searched for it and eventually found it near the Agrigentine gate in Syracuse, in a neglected condition and overgrown with bushes. Cicero had the place tidied up, and it was said that he could see the carving and read some of the verses that had been added as an inscription. But the tomb was again lost as time went by. There is a claim that the tomb discovered in the courtyard of the Hotel Panorama in Syracuse in the early 1960s was Archimedes'; unfortunately there is no compelling evidence for this and many historians consider that the location of his tomb today is unknown [2].

Fig. 4.3: Death of Archimedes at the hands of a Roman soldier, as depicted in a sixteenth century mosaic by an unknown artist [9]. The Roman fleet, commanded by Marcellus, laid siege to Syracuse and the city fell after three years of war. Archimedes was killed (212 BC) by a Roman soldier during the sack of the city. It is believed that Marcellus honoured [4] Archimedes with a proper funeral and erected a tombstone as Archimedes had desired.

The engineering tradition started by Archimedes influenced several contemporaries and future generations. One among them, who lived during Archimedes's time, was Ctesibius ($\sim$ 285–222 BC), who devised a practical water clock. He improved upon the more ancient Egyptian device (called the *clepsydra*), in which water dripped into a container at a steady rate, thereby moving a pointer that would indicate the time. Ctesibius made his device practical, compact, and accurate. In fact, his clock was as accurate as those of the Middle Ages, which were run by falling weights. Accuracy in the measurement of time only improved significantly after the invention of the pendulum (see Chap. 19).

Box 4.2: Archimedes' War Machines

One of the most dramatic tales about Archimedes comes to us from the description of Marcellus in the works of Plutarch, in the context of the siege of Syracuse.

Hieron II, King of Syracuse, had a treaty of alliance with Rome. After his death, his grandson Hieronymus occupied the throne of Syracuse. During his reign, Rome suffered a major defeat in its war with Carthage and seemed to have been thoroughly beaten. Hieronymus, misjudging the situation, switched his loyalties to the winning side, Carthage. The Romans, of course, didn't like this at all, and once they recovered, they sent a fleet commanded by Marcellus, to lay siege to Syracuse.

This, if Plutarch is to be believed, started the strange three-year war between the mighty Roman fleet on one side and a one-man-army, viz., Archimedes, on the other side. The mechanical inventions Archimedes used in this war probably constituted the first massive application of superior technological knowledge in warfare. He is said to have constructed large mirrors and lenses to set Roman ships on fire and huge mechanical cranes to lift ships from the sea (Fig. 4.4). This account was originally given by Polybius in his *Universal History* [5], written about seventy years after Archimedes' death and used as a source for Plutarch.

There is, however, some amount of controversy about whether Archimedes really did all this! The main reason is that no one has been able to reconstruct mirror systems that would actually cause serious damage to ships (compared to other weapons available at that time) [4].

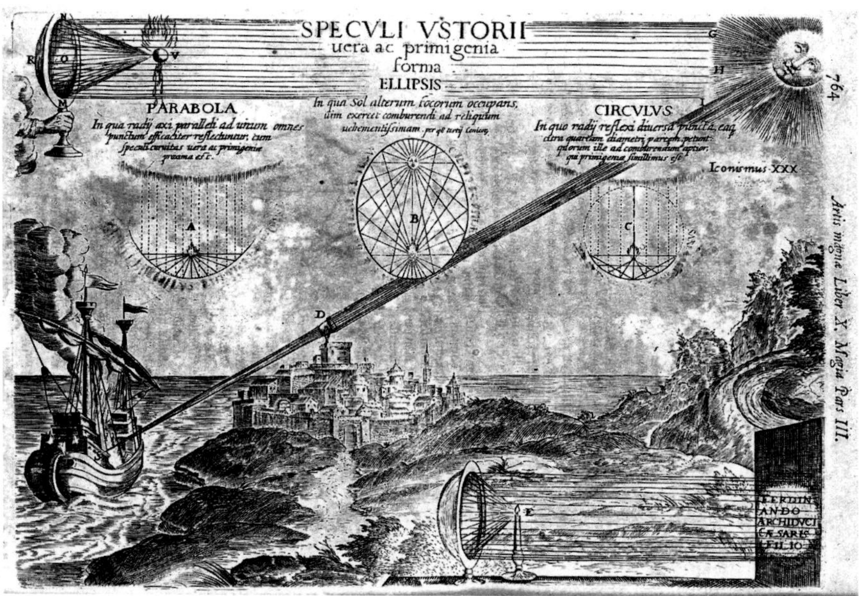

Fig. 4.4: This seventeenth century engraving shows the arrangement Archimedes is supposed to have used to burn Roman ships, during the siege of Syracuse [10], using large mirrors and lenses. It is also claimed that he built huge mechanical cranes to lift ships from the sea. These reports are from the description of Marcellus in the works of Plutarch, but there is a fair amount of controversy as to whether Archimedes actually did any of this.

After about 120 BC, Ptolemaic Egypt fell into decadence, and by 30 BC it had become a Roman Province. Greek science was virtually over but for an occasional genius. One such man was Hero of Alexandria (10–70 AD). His most famous invention was a hollow sphere with two tubes attached to it; when the water in the sphere was boiled to make steam, the steam escaping through the tubes made the whole device spin. This was indeed the first steam engine, though — unfortunately — it was only used in toys and by priests to deceive gullible believers. Hero also wrote extensively on mechanics, elaborating on the principle of the lever and discussing several simple machines involving inclined planes, pulleys, and levers.

Our knowledge of Archimedes and his work increased significantly in recent times when historians of mathematics discovered fresh information in the form of an ancient parchment which had been overwritten by monks nearly a thousand years ago. Such documents, called palimpsests, were created by scraping the ink

Fig. 4.5: One page of the Archimedes Palimpsest from his work *On Floating Bodies* [11]. One common writing medium used in ancient times was specially prepared animal skin. (There is an interesting story about how this practice started; see Box 8.1.) Since this medium was a rather precious commodity, it was customary to re-use it by scraping the ink off the animal skin and entering new text. Such documents were called palimpsests. In 1906, the Danish professor Johan Ludvig Heiberg (1854–1928) came across a 174-page goatskin parchment in Constantinople, mainly containing prayers. On closer examination, it turned out that the prayers were overwritten on a previous work describing the contributions of Archimedes! Indeed, the earlier entry dated back to the tenth century AD and contained previously unknown treatises by Archimedes [6]. The parchment, which had been in a monastery library in Constantinople for hundreds of years, was later sold to a a private collector in the 1920s. It hit the newpaper headlines when, on 29 October 1998, Christie's in New York auctioned it to an anonymous buyer for 2 million dollars! It is currently under study at the Walters Art Museum in Baltimore.

off existing works and reusing them, a common practice in the Middle Ages when vellum — specially prepared animal skin used as a medium for writing — was in short supply (see also Box 8.1).

In 1906, a Danish professor, Johan Ludvig Heiberg (1854–1928) visited Constantinople and examined a 174-page goatskin parchment containing prayers

Fig. 4.6: The highest honour in mathematics, the Fields Medal, is often taken to be the equivalent of the Nobel Prize [12]. Befittingly, it carries the portrait of Archimedes on one side of the medal. The inscription around his head contains a quote which is attributed to him: "Transire suum pectus mundoque potiri", meaning "Rise above oneself and grasp the world". The other side of the medal shows a sphere embedded in a cylinder, Archimedes' favourite diagram.

originally written in the thirteenth century AD. He discovered, to his surprise, that the older works in the palimpsest were tenth century AD copies of previously unknown treatises by Archimedes. The parchment had been languishing for hundreds of years in a monastery library in Constantinople before it was sold to a private collector in the 1920s. On 29 October 1998, it was sold at an auction, at Christie's in New York, to an anonymous buyer for 2 million dollars — a sale which hit newspaper headlines! The palimpsest is currently kept at the Walters Art Museum in Baltimore, Maryland, where it has been subjected to a range of modern tests to read the overwritten text.

This palimpsest holds seven treatises, including what appears to be the only surviving copy of *On Floating Bodies* in the original Greek (Fig. 4.5). This parchment gives us information, among other things [6], about a curious puzzle called the stomachion. The origin of this term is unclear, but it involves fairly advanced concepts from the branch of mathematics now called combinatorics. The goal of the stomachion puzzle is to determine in how many ways a particular set of 14 pieces of varied planar figures can be put together to form a square. In 2003, mathematicians found that the answer is actually 17 152!

The Fields Medal, considered to be the equivalent of the Nobel Prize in mathematics, honours Archimedes by having his profile on the obverse side of the medal. The inscription around his head is a quote attributed to him: "Transire suum pectus mundoque potiri", meaning, "Rise above oneself and grasp the world" (Fig. 4.6).

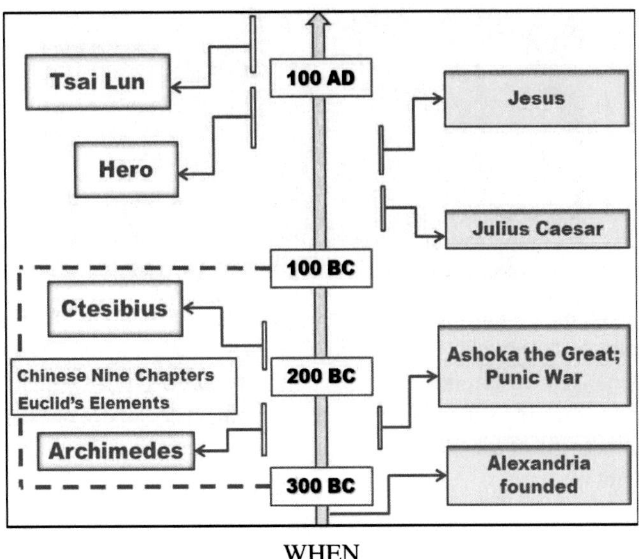

WHEN

Notes, References, and Credits

Notes and References

1. For more on Archimedes, see, e.g.,
   Stein, Sherman (1999), *Archimedes: What Did He Do Besides Cry Eureka?* Mathematical Association of America, Washington DC [ISBN: 978-0-88385-718-2].
   Clagett, Marshall (1964–1984), *Archimedes in the Middle Ages* (5 Volumes), University of Wisconsin Press, Madison.
   Dijksterhuis, E.J. (1987), *Archimedes*, Princeton University Press, Princeton [ISBN 0-691-08421-1].
   Gow, Mary (2005), *Archimedes: Mathematical Genius of the Ancient World*, Enslow Publishers, Inc., New Jersey [ISBN 0-7660-2502-0].
   Hasan, Heather (2005), *Archimedes: The Father of Mathematics*, Rosen Central, US [ISBN 978-1-4042-0774-5].
2. See e.g., Rorres, Chris, *Tomb of Archimedes: Sources*
   http://www.math.nyu.edu/∼crorres/Archimedes/Tomb/Cicero.html.
   Courant Institute of Mathematical Sciences Archive
   (https://web.archive.org/web/20061209201723/http://www.math.nyu.edu/∼crorres/Archimedes/Tomb/Cicero.html).
3. For more on the history of π, see, e.g.,

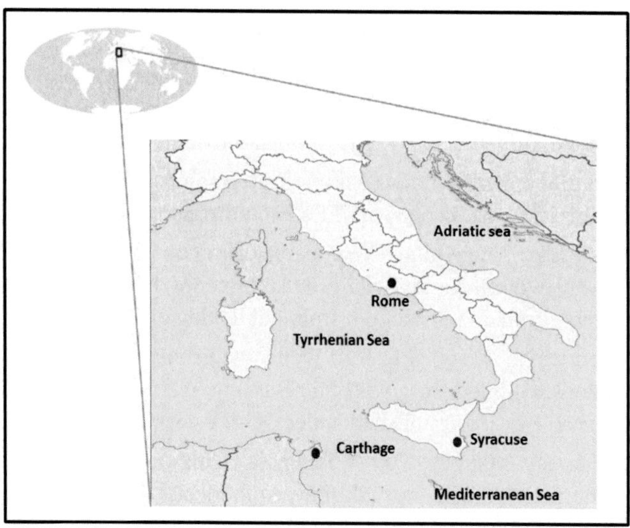

WHERE

Beckmann, Peter (1989), *History of Pi*, St. Martin's Press, New York [ISBN 978-0-88029-418-8].

Blatner, David (1999), *The Joy of Pi*, Walker and Company, US [ISBN 978-0-8027-7562-7].

Zebrowski, Ernest (1999), *A History of the Circle: Mathematical Reasoning and the Physical Universe*, Rutgers University Press, New Jersey [ISBN 978-0-8135-2898-4].

4. See, e.g.,
Russo, Lucio (2013), *Archimedes between legend and fact*, Lett. Mat. Int. **1**, 91–95.
Mills A.A and Clift, R. (1992), *Reflections of the Burning mirrors of Archimedes with a consideration of the geometry and intensity of sunlight reflected from plane mirrors*, Eur. J. Phys. **13**, 268.
Simms, D.L (1975), *Archimedes and burning mirrors*, Physics Education **10**, 517
as well as the website: https://explorable.com/archimedes-war-machines
5. For an English version, see:
https://www.math.nyu.edu/~crorres/Archimedes/Siege/Polybius.html
6. Netz, Reviel and Noel, William (2007), *The Archimedes Codex*, Orion Publishing Group, UK [ISBN 0-297-64547-1].

Figure Credits

7. Figure 4.1 courtesy: Benjamin West /Wikimedia Commons / Public Domain; https://commons.wikimedia.org/wiki/File:West,_Benjamin_-_Cicero_Discovering_the_Tomb_of_Archimedes_1797.jpg (from public domain).
8. Figure 4.2 courtesy: Anonymous /Wikimedia Commons / Public Domain; https://commons.wikimedia.org/wiki/File:Archimede_bain.jpg (from public domain).
9. Figure 4.3 courtesy: Anonymous /Wikimedia Commons / Public Domain; https://commons.wikimedia.org/wiki/File:Archimedes_before_his_death_with_the_Roman_soldier,_Roman_mosaic.jpg (from public domain).
10. Figure 4.4 courtesy: L0016024 Credit: Wellcome Library, London; https://commons.wikimedia.org/wiki/File:Burning_mirrors,_from_Kircher._Wellcome_L0016024.jpg (published under CC-BY-4.0).
11. Figure 4.5 courtesy: Matthew Kon /Wikimedia Commons / Public Domain; https://commons.wikimedia.org/wiki/File:Archimedes_Palimpsest.jpg (from public domain).
12. Figure 4.6 courtesy: Stefan Zachow /Wikimedia Commons / Public Domain; https://commons.wikimedia.org/wiki/File:FieldsMedalFrontAndBack.jpg (from public domain).

5

The Healing Art and Its Science

When they used a variety of plants and plant products for food, primitive tribes must have realized that some of these were poisonous and some were curative. Those in the tribe who were clever enough to appreciate these effects could assert considerable power over others. They were the earliest 'witch doctors', attributed magical powers. Healing soon got inextricably mixed up with magical practices, gods, and demons. And medicine was in the hands of god-men and witch doctors (and it remains so even today, although to a much lesser extent!). To retain their special status, it was also necessary for early medicine men to shroud the details of their practices in mystery.

It is, therefore, rather surprising that the science of medicine *did* develop to a high degree in two of the ancient civilizations — the Indian and the Greek — in spite of their abiding supernatural beliefs. The earliest concepts of Indian medicine are presented in one of the four Vedas — the sacred Indian texts — called the Atharvaveda, which probably dates back to 2000 BC. Several common diseases like fever, coughs, diarrhea, abscesses, etc., and their herbal remedies are mentioned in the Vedas. Unfortunately, the cures are mixed up with spurious magical practices making it difficult to perform an objective evaluation.

The golden age of Indian medicine, however, occurred in the post-Vedic period, sometime during 800 BC to AD 100. Two major medical treatises — the Charaka Samhita [1] and the Sushruta Samhita [2] — appeared during this period. These texts discuss in detail several aspects of medicine: symptoms, diagnosis, and

T. Padmanabhan and V. Padmanabhan, *The Dawn of Science*,
https://doi.org/10.1007/978-3-030-17509-2_5

classification of diseases, preparation of medicines from plants, diet, and care of patients, etc.

Charaka, who probably lived between the sixth and second century BC, is known as the "Father of Indian Medicine". He was one of the key contributors to Ayurveda, the system of medicine (and lifestyle for healthy living) developed in ancient India. He authored an influential text (possibly before the second century, though this date is somewhat uncertain) called the Charaka Samhita [3]. (An earlier scholar, Agnivesa, had written an encyclopedic treatise in the eighth century B.C. However, only after a comprehensive revision by Charaka did this treatise gain popularity, later to become known as Charaka Samhita.) Charaka Samhita consists of eight books and one hundred and twenty chapters dealing with the basic principles of Ayurveda practice.

Fig. 5.1: Ancient Indian medicine achieved the most in the field of surgery [7]. In his works (produced around AD 100), the Indian scholar, Sushruta, gave very detailed instructions about different kinds of surgical operations, such as the excision of tumors, incision of abscesses, stitching of open wounds, etc., and about the proper choice of instruments for each of these procedures. These instruments include knives of different kinds, scissors, trocars (for piercing tissues), saws, etc., and were mostly made of steel.

Ayurveda was based on the idea that all diseases are caused by an imbalance among three vital entities acting in the body: air (vayu or vata), phlegm (kapha), and bile (pitta). The seven constituents of the body — taken to be blood, flesh, fat, bone, marrow, chyle, and semen — were supposed to be produced by the action of these three entities. Most of the cures involved restoring the balance between these three entities by dietary and herbal means. Charaka lists nearly 500 medicinal plants, while Sushruta has a more extensive list of 760. In addition, several animal products and minerals were also used in the treatments. The Indian medical man could administer emetics, purgatives, enemas, and also sneezing powders and herbal fumes as part of a treatment.

Charaka Samhita is unique in emphasising a rational approach to the diagnosis and cure of diseases. Charaka also gives a predominant place to methods of prevention of all types of diseases, including restructuring one's lifestyle to be in harmony with the course of nature and the four seasons. For nearly two millennia, this work remained a standard reference on the subject and it was translated into Arabic and Latin during the Middle Ages.

While all these were important advances in medical knowledge, it was in the field of surgery that Indian medicine achieved the most. By around AD 100, several surgical procedures were known and practised; these included the excision of tumors, incision of abscesses, removal of fluids from parts of the body, probing of fistulas, and stitching of open wounds (Fig. 5.1). The classical texts give very detailed instructions about these operations and about the proper choice of instruments. Sushruta, for instance, describes 20 sharp instruments and about 100 non-cutting ones — including knives of different kinds, scissors, trocars (for piercing tissues), saws, needles, forceps, levers, and hooks. Most of these instruments were made of steel and the operations seem to have been performed using alcoholic products as anesthetics.

Almost around the same time, the art of healing was thriving in Greece as well. Hippocrates (460–370 BC) seems to have been the first person to state categorically that diseases are due to natural causes and not a consequence of curses from gods — a major breakthrough in those days! Very little is known about his life and work [4]. Some historians think that most of the works attributed to him could have been written by several other people who lived much later. (Since Hippocrates's name carried considerable weight in Greece, people probably preferred to attribute their own ideas to Hippocrates!). Whatever the truth may be, the books that make up the collection, called *Corpus Hippocraticum*, earned him the title of 'father of modern medicine'. In this work, he describes the symptoms and the courses

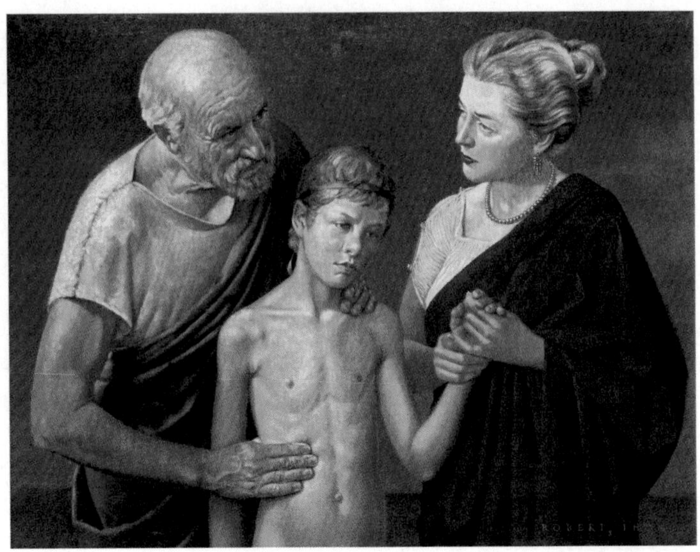

Fig. 5.2: Hippocrates (460–370 BC) is widely considered to be the Father of Medicine in the West. The major conceptual advance made by him was in realizing — and emphasising — the fact that diseases arise due to natural causes and not because of curses from the gods! The set of books attributed to him (though historians doubt whether he authored all of them!), called *Corpus Hippocraticum*, clearly describes the symptoms and the courses of several illnesses. He also developed a code of conduct for medical practitioners, which is still used for medical graduates all over the world, in a form known as the 'Hippocratic Oath' [8].

of several illnesses clearly and concisely. Hippocrates repeatedly emphasized the natural causes for the illnesses and sought to cure them in a methodical way.

For example, *Corpus Hippocraticum* contains a treatise on epilepsy, which was widely considered to be due to 'possession' and was even called the 'sacred disease', because supernatural beings were supposed to have been involved. In the book *On the Sacred Disease*, written around 400 BC (possibly by Hippocrates himself, though authorship of this book, unfortunately, is not confirmed), this view is strongly refuted. Hippocrates refused to attribute divine causes to *any* disease, and epilepsy was no exception. Like all other diseases, he expected epilepsy to have a natural cause and a rational treatment. Even if the cause was not known, and the treatment was unclear, Hippocrates argued, the principle remained the same. Possibly, the birth of the science of biology and medicine could be identified with the creation of this book, *On the Sacred Disease*, around 400 B.C.

Hippocrates also laid a fair amount of emphasis on the effects of diet, occupation, climate, and environment on health, which sounds almost contemporary! His greatest legacy perhaps was in developing the code of conduct for medical practitioners — known as the 'Hippocratic Oath' — which is used even today at medical graduation ceremonies all over the world.

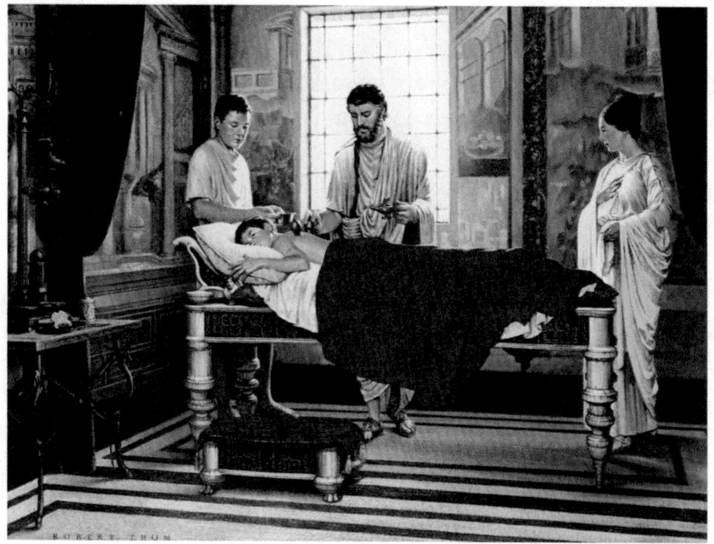

Fig. 5.3: Galen's ($\sim$ 129–200 AD) work, written in Rome around 140 AD, remained the most influential doctrine in Europe until the sixteenth century. This painting shows him using the technique called 'cupping', that is, creating small vacua in heated cups to 'draw the poisons out', a procedure which continued to be practised in folk culture right up to the early twentieth century. Galen recognized that the arteries carried blood, and exhibited rhythmic motion in synchrony with the beating of the heart. In fact, he narrowly missed discovering the circulation of blood [9]!

The first formal medical school was established (again!) in Alexandria [5] around 300 BC and thrived under the Greek anatomist, Herophilus (335–280 BC). He was the first to dissect human bodies in public, breaking existing traditions.

(In the pre-Christian era, there was no taboo on dissection and the Greeks took full advantage of it. In contrast, ancient Indians desisted from *cutting* open bodies. Hence, Sushruta Samhita recommended an elaborate procedure for soaking a body in water so that parts could be removed without cutting!) Herophilus gave detailed descriptions of the brain, parts of the eye, ovaries, and uterus; he also identified and named the retina, the duodenum, and the prostate gland. These investigations were continued later on by Erasistratus (304–250 BC), who also taught at Alexandria. By now, medical science had detailed knowledge of the various organs of the body, though it had very little idea of their functions.

Soon after Erasistratus, the study of anatomy declined in the West because of religious objections to dissection of the human body. All later workers, notably Galen ($\sim$ 129–200 AD), had to rely on animal dissections to understand anatomy. In spite of this constraint, Galen could make progress. He was probably the first to notice that arteries carried blood, and were set into a rhythmic motion by the pounding of the heart. He also used the pulse as a diagnostic test but, in spite of all this, failed to discover the circulation of the blood (see also Chap. 15).

Box 5.1: Chinese Medicine

Chinese medicine has great antiquity. Several medical texts originated in China during the period 500 BC to AD 300, of which, the most famous were the compilations Nei jing and Mo Ching. Huangdi Nei jing ("Inner Canon of the Yellow Emperor" or "Esoteric Scripture of the Yellow Emperor"), is an ancient Chinese medical text that has been considered the basic source for ancient Chinese medicine for more than two millennia. This comprises two texts — each of eighty-one chapters presented in a question-and-answer form between the mythical Yellow Emperor and six of his legendary ministers. The first text, the *Suwen*, (sometimes known as "Basic Questions"), covers the theoretical foundations of medicine and diagnostic methods. The second text discusses acupuncture therapy in great detail [6].

Traditional medical practice in China was based on the theory of Yin and Yang, which were taken to be the two fundamental principles of Nature. Illnesses are supposed to be caused by the imbalance between Yin and Yang in the body. Unfortunately, the Chinese understanding of the anatomy was primitive — again because of religious restrictions on the dissection of the body — but they made good progress in diagnostics. Mo Ching, for example, describes detailed guidelines for the interpretation of the pulse, which is to be measured not only at the wrist but also at certain other parts of the body!

On the other hand, acupuncture is a mysterious — and totally Chinese — concept in medicine, which is actively practised even today. This consists in inserting hot or cold metal needles — ranging in lengths from 3 to 25 cm — into the body at various points. Practitioners claim that these needles provide the cure, by changing the distribution of Yin and Yang in the body and restoring the balance between them. Acupuncture dates from before 2500 BC and is significantly different from the more widely practised concepts of modern (Western) medicine.

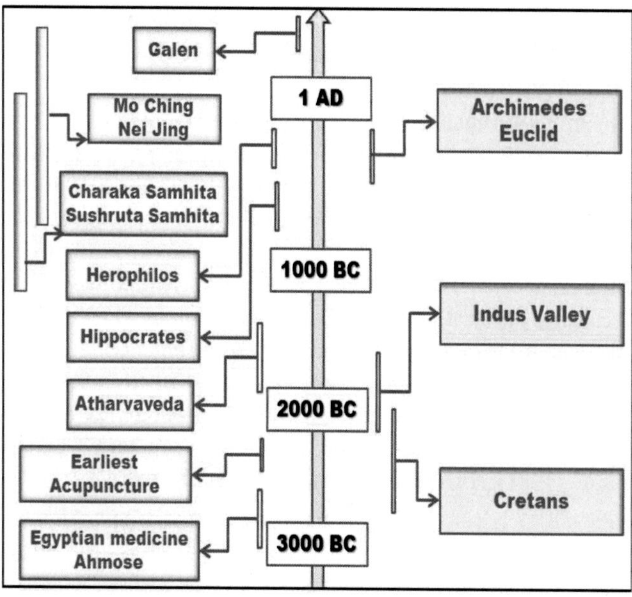

WHEN

Notes, References, and Credits

Notes and References

1. For a description of Charaka Samhita and related topics see:
   Valiathan, M. S. (2003), *The Legacy of Caraka*, Orient Longman, India [ISBN 81-250-2505-7].
   Robert Svoboda (1992), *Ayurveda: Life, Health and Longevity*, Penguin Books pp. 189–190 [ISBN 978- 0140193220].
   Valiathan, M.S (2009), *An Ayurvedic view of life*, Current Science, Volume **96**, Issue 9, 1186–1192.
   Kaviratna, Avinash C. and Sharma, P. (1913), *The Charaka Samhita* (5 Vols), Sri Satguru Publications, New Delhi [ISBN 81-7030-471-7].
   Wujastyk, Dominik (2003), *The Roots of Ayurveda*, Penguin Classics (3rd edition) [ISBN 978-0140448245], pp. 1–50 gives an introduction to the Charaka Samhita and a modern translation of selected passages.
2. More on Sushruta Samhita and related topics can be found in:
   Loukas, Marios et al. (2010), *Anatomy in ancient India: a focus on the Susruta Samhita* (review), J. Anat. **217**, 646–650 [doi: 10.1111/j.1469-7580.2010.01294.x].
   Kutumbian, P (2005), *Ancient Indian Medicine*, Orient Longman, India [ISBN 978-812501521-5].
   Ray, Priyadaranjan et al. (1980), *Susruta samhita: a scientific synopsis*, Indian National Science Academy Publications, New Delhi, number OCLC 7952879.
   Meulenbeld, Gerrit Jan (1999), *A History of Indian Medical Literature*, Brill Academic Publishers, Groningen, (all volumes, 1999–2002) [ISBN 978-9069801247].
   Sharma, P. V. (1992), *History of medicine in India, from antiquity to 1000 A.D*, Indian National Science Academy publication, New Delhi, number OCLC 26881970.
   Schultheisz, E. (1981), *History of Physiology*, Pergamon Press [ISBN 978-0080273426].
3. See, Meulenbeld, Gerrit Jan (1999), *A History of Indian Medical Literature*, E. Forsten, Groningen [ISBN 978-90-6980-124-7]. Volume IA, pp. 7–180 of this work gives a detailed survey of the contents of the Charaka Samhita and a comprehensive discussion of all historical matters related to the text, its commentators, and its later history in the Islamic world and in Tibet.
4. More on Hippocrates:
   Pinault, Jody Robin (1992), *Hippocratic Lives and Legends*, Leiden: Brill Academic Publishers, Groningen [ISBN 90-04-09574-8].
   Smith, Wesley D (1979), *Hippocratic Tradition*, Cornell University Press, New York [ISBN 0-8014-1209-9].

Edelstein, Ludwig (1996), *The Hippocratic Oath: Text, Translation, and Interpretation*, Johns Hopkins University Press, Baltimore [ISBN 978-0801801846].
Goldberg, Herbert S. (1963), *Hippocrates, Father of Medicine*, Franklin Watts, New York [ISBN 978-0531008836].
Heidel, William Arthur (1941), *Hippocratic Medicine: Its Spirit and Method*, Columbia University Press, New York.
Margotta, Roberto (1968), *The Story of Medicine*, Golden Press, New York.
Marti-Ibanez, Felix (1961), *A Prelude to Medical History*, MD Publications Inc, New York.

5. More on Herophilus:
Lloyd, G. E. R. (1973), *Greek science after Aristotle*, Norton, New York [ISBN 0-393-04371-1].
Lloyd, G. E. R. (1983), *Science, folklore and ideology: studies in the life sciences in ancient Greece*, Cambridge University Press, Cambridge [ISBN 0-521-25314-4].
6. For an English translation of this text, see:
Unschuld P. U. and H. Tessenow (2011), *Huang Di nei jing su wen. An Annotated Translation of Huang Di's Inner Classic – Basic Questions*, University of California Press, Berkeley, US [978-0520266988].

Figure Credits

7. Figure 5.1 is from "The History of Medicine, 1952". Reproduced here with permission from the Collection of Michigan Medicine, University of Michigan, Gift of Pfizer, UMHS.6.
8. Figure 5.2 is from "The History of Medicine, 1952". Reproduced here with permission from the Collection of Michigan Medicine, University of Michigan, Gift of Pfizer, UMHS.7.
9. Figure 5.3 is from "The History of Medicine, 1952". Reproduced here with permission from the Collection of Michigan Medicine, University of Michigan, Gift of Pfizer, UMHS.8.

6

The Legacy from the Arab World

In the declining years of the third century AD, the Roman empire was split into two. Of these, the eastern part flourished well and emerged as the Byzantine empire in the course of time; but the western part drifted down till it fell to the barbarian conquests by the end of the fifth century. With that began many years of instability and anarchy in western Europe. The stage was set for the world to lose forever the earlier scientific contributions of the Greeks, which would have been a disaster. And indeed, that is what would have happened — but for the phenomenal rise of the Arab civilization, which preserved the earlier knowledge of humanity.

In the years following the death of Prophet Mohammed in 632 AD, the Arabs rapidly conquered Asia Minor, Persia, North Africa, and Spain. During the ninth to the eleventh centuries, the Arab civilization was the dominant influence around the Mediterranean with Cordoba, Baghdad, Damascus, and Samarkand emerging as centres of learning and culture. *Corboda had a library of 400 000 books and its streets were paved and lit by lamps — amenities that were not available in London or Paris for another seven centuries!*

The Arabs absorbed every key idea of Greek science and added to it what they learnt from Persia, India, and China. The Caliphs in Baghdad encouraged the translation of every major scientific work into Arabic. The so-called 'western science' of later centuries originated from the study of these Arabic texts, re-translated into Latin. This Arabic reservoir of knowledge covered virtually all branches of science developed in antiquity — medicine, chemistry, astronomy, mathematics, and physics.

T. Padmanabhan and V. Padmanabhan, *The Dawn of Science*,
https://doi.org/10.1007/978-3-030-17509-2_6

Fig. 6.1: Left: A miniature portrait [9] of Avicenna (980–1037 AD), a child prodigy who contributed extensively to medicine in the Arab world. He wrote more than a hundred books, many of which were so authoritative that their Latin translations were used in Europe even as late as 1650 AD! Right: Photograph showing a monument to Avicenna in Dushanbe [10].

In the field of medicine, two names stand out from the Arab world [1] — Al-Rhazes (854–925 AD) and Avicenna (980–1037 AD). Rhazes was born near Tehran and studied medicine in a school in Baghdad. In addition to contributing his share to the translations of texts, he also wrote a voluminous treatise on medicine (Fig. 6.3), *Kitab-al-hawi*, covering the subject in its entirety. Rhazes [2] was also an excellent chemist; he was the first person to make what we now call 'plaster of Paris', using it to set broken bones.

Avicenna, the son of a wealthy tax collector, was a child prodigy who also had the extra advantage of the best education money could buy at the time. During his career, he wrote more than a hundred books, many of which were on medicine. His works were so authoritative that the Latin translations of these books were used in Europe even as late as 1650 AD! His treatise, *Canon of Medicine* (Fig. 6.3), was the standard medical textbook in Europe till the seventeenth century (and even later) and is still used as an authoritative source by some traditional healers in the Middle East. The work consists of five books, which includes a general discussion of the

Fig. 6.2: European depiction [11] of the Persian doctor Al-Rhazes (854–925 AD), in Gerardus Cremonensis "Recueil des traités de médecine" ($\sim$ 1250–1260 AD). He was a student of the school of medicine in Baghdad and the author of a major treatise on medicine, *Kitab-al-hawi*. Rhazes [2] was also an excellent chemist and was the first person to make — what we now call — 'plaster of Paris'. He used it for making a cast around broken bones!

scientific background to medicine and anatomy, a discussion of the therapeutic properties of substances used in medicine, a book pertaining to specific, localized ailments, and another to general diseases that affect the whole body (like fever, etc.), and finally a treatise on pharmacology. Avicenna based his system of medicine on the Graeco-Roman tradition, in particular that of Galen ($\sim$ 129–200 AD).

Another notable Arab scholar of this period was Ibn al-Nafis (12131288 AD), who wrote a detailed commentary on the Anatomy in Avicenna's *Canon of Medicine*. In this commentary, he gives a clear description of blood circulation which — as we shall see in Chap. 15 — was rediscovered several times, ending with William Harvey (1578–1657 AD)! In addition to this key contribution, Ibn al-Nafis also authored — what is considered to be — the largest medical encyclopedia ever written by one person. Called *Al-Shamil fi al-Tibb* (The Comprehensive Book on

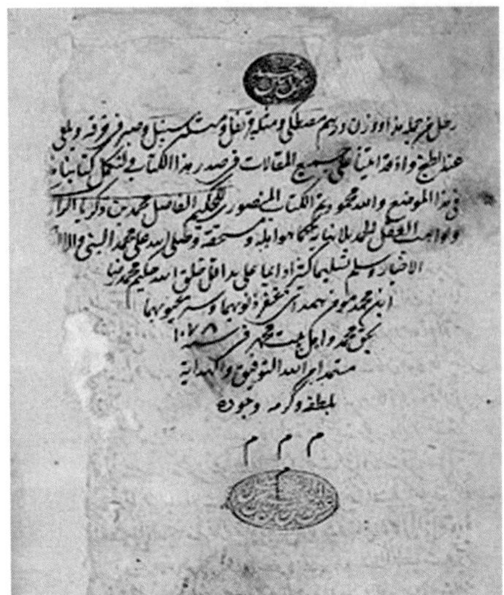

Fig. 6.3: Left: Colophon of Al-Rhazes's Book of Medicine[12]. Right: An Arabic copy of the *Canon of Medicine*, Vol. 5 by Avicenna, dated 1052 AD (Aga Khan Museum, Toronto) [13]. These works exerted great influence in the field of medicine for centuries.

Medicine), it was planned to comprise 300 volumes; but Ibn al-Nafis managed to publish only 80 of them before his death. All the same, it gave a complete summary of all the medical knowledge in the Islamic world at the time. Ibn-al-Nafis bequeathed his encyclopedia, along with the rest of his library, to the Mansoory hospital, where he was working before his death. Over a span of time, many of the volumes of the encyclopedia got lost or dispersed in different parts of the world. Currently only 2 volumes are still around in Egypt [3].

Another important Arab contribution to medicine was in the actual preparation of drugs. Many drugs we use today are of Arabic origin and so also are processes like distillation and sublimation. The Arabs, in collaboration with Nestorian Christians (an Eastern Church that was not affiliated with Constantinople), established a major hospital at Jundi Shahpur (in south-west Persia), which combined under one roof the treatment of patients, medical education, and the translation of medical texts; such a combination of academic and clinical activity was the first of its kind.

In contrast to medicine, the branch of science we now call chemistry was rather weak in ancient Greece and also in the Arab world. The origins of this discipline can be found in the ancient Egyptian science called 'khem', which was used to preserve mummies. Very little is known of the original Egyptian work on this subject. Most of the information comes from a 28-volume encyclopedia authored by the Greek scientist, Zosimus (around 300 AD); unfortunately, the contents of these volumes are soaked in deep mysticism.

The primary interest of most of the early and medieval alchemists (al' is the Arabic word for 'the'; the word 'chem' came from the Egyptian 'khem') was to discover a method to convert base metals into gold. Medieval history is full of names of scientists who performed, on the one hand, valuable experiments and made accurate observations, but who, on the other hand, also wasted a lot of their time looking for a mysterious substance, al-iksir, which would transmute base metals into gold (al-iksir later became 'elixir' in Latin).

One such scientist was Geber (721–815 AD) (also known as Abu Musa Jabir ibn Hayyan) [4], who lived at the same time as Haroun-al-Rashid of Arabian Nights fame. He successfully produced several chemical compounds like white lead, ammonium chloride, nitric acid, and acetic acid but was (probably) really hoping to find the al-iksir! Unfortunately, later alchemists followed Geber's wrong theories into wilder morasses, while ignoring his really important contributions. For all these meanderings, somewhat amusingly, the Arabs excelled in practical 'everyday' chemistry. They were, for example, pioneers in making perfumes. In fact, perfume-making was practised as a household art in upper class Arab families.

The Arabs were also instrumental in transmitting two major innovations in chemical technology from the Orient to the West — the manufacture of paper and gunpowder.

Paper was known in China as early as the first century AD. Almost at the same time as Hero was devising his machines (see Chap. 4), a Chinese inventor named Tsai Lun (~ 48–121 AD), also written as Cai Lun, made a breakthrough in China. Chinese historians credit him with inventing the product [5] which we now call 'paper' from tree bark and rags (Fig. 6.4) around the year 105 AD. The official history of the Han dynasty (third century AD) tells us that Tsai Lun presented Emperor He of Han (who ruled from 88–106 AD), with samples of paper and was given an aristocratic title and riches. (It did not last; it turns out that Tsai Lun was also involved in a palace murder and Emperor An of Han, who ruled from 106–125 AD, ordered his imprisonment. Tsai Lun committed suicide by taking poison.)

Over the next few centuries, paper-making spread westwards. The Arabs actually captured some of the Chinese craftsmen in the battle of Samarkand (751 AD) and learnt the art from them. Baghdad had perfected this technique by 800 AD, and within the next 50 years, the first Arab paper-makers were plying their new trade

Fig. 6.4: Chinese drawing describing the procedures used to make paper in ancient China [14] . The art of paper-making was known in China as early as the first century AD. The Arabs learnt about this from some Chinese craftsmen, captured by them in the battle of Samarkand (751 AD). Within the next 50 years, the first Arab paper-makers had started their full-scale trade in Baghdad. But it took another four centuries for the art to make any significant appearance in Europe.

in Baghdad. The penetration of paper making into Europe, however, was a slow process, taking another four centuries; Europe inherited it only after the Crusades (in the thirteenth century). In the twenty centuries since Tsai Lun, this invention has yet to be improved significantly!

The story of gunpowder is less clear [6]. The discovery of gunpowder in ancient China during the Han period (202 BC–220 AD) was probably an accident, arising from alchemy. Chinese alchemists — like others — were trying to make gold or prepare an elixir of immortality. They used sulfur and saltpeter, which takes one close to gunpowder. In fact, a book of the late Tang dynasty (but probably containing material from much earlier dates), explicitly cautions alchemists to take due care when dealing with sulfur, saltpeter, and charcoal.

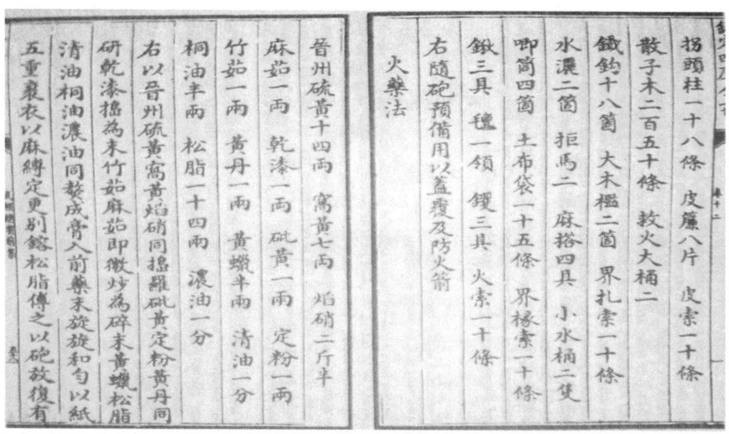

拐頭柱一十八條 皮簾八片 皮索一十條
散子木二百五十條 救火大桶二
鐵鈎十八箇 大木檑二箇 界扎索一十條
水濃二箇 拒馬二 麻搭四具 小水桶二隻
唧筒四箇 土布袋一十五條 界椽索一十條
鍬三具 钁一領 鏵三具 火索一十條
右隨砲預備用以蓋覆及防火箭
火藥法
晉州硫黃十四兩 窩黃七兩 焰硝二斤半
麻茹一兩 乾漆一兩 砒黃一兩 定粉一兩
竹茹一兩 黃丹一兩 黃蠟半兩 清油一分
桐油半兩 松脂一十四兩 濃油一分
右以晉州硫黃窩黃焰硝同搗羅砒黃定粉黃丹同
研乾漆搗為末竹茹麻茹即微炒為碎末黃蠟松脂
清油桐油濃油同熬成膏入前藥末旋旋和勻以紙
五重裹衣以麻縛定更別鎔松脂傅之以砲放復有

Fig. 6.5: These pages [15] contain the earliest known written formula for gunpowder; this is from the Chinese, Wujing Zongyao of 1044 AD. Gunpowder, invented by the Chinese, was originally used only for ornamental fireworks. The Arabs probably learnt this technique as early as 800 AD, but its full-scale use in European weaponry may only have started in the thirteenth century.

The earliest explicit reference to gunpowder is from the Eastern Han dynasty around 142 AD. The alchemist Wei Boyang belonging to this period had written about a substance with properties very similar to gunpowder. In his work, *Cantong qi*, also known as the Book of the Kinship of Three (a Taoist text on the subject of alchemy), he had described a mixture of three powders that would 'fly and dance' violently. In China, gunpowder (Fig. 6.5) was orginally used for entertainment purposes and during religious and ceremonial functions.

There is no record of the use of gunpowder in the battlefield before the tenth century. In the early twelfth century, the knowledge of gunpowder was passed on to the Jurchens, who established the Jurchen empire (1115–1234 AD) in northern China. When they were defeated by the Mongols in 1232–1233 AD, both their gunpowder and firearm technicians were captured. Later, when they overran Europe, the Mongols deployed firearms in their offensive.

Meanwhile, the Arabs had set up the Mamluk Caliphate in Cairo and, in 1260, they defeated the Mongols in a battle in Syria. Once again, the firearms and the technicians were acquired by the victorious side — by the Arabs from the captured Mongols. The Arabs introduced them into the battlefield and used them successfully in the Crusades later on, for example. Europe learnt to use them over the next few centuries.

Box 6.1: Greek Fire

Though chemistry was never one of the Greeks' strong points, they did show remarkable ingenuity in this field in times of need. One such invention was the chemical mixture called 'Greek fire', attributed to the seventh century alchemist, Callinicus, of Greek origin.

Callinicus is supposed to have fled Syria to Constantinople ahead of the invading Arabian armies, and here he invented the Greek fire to fight the Arabs. This mixture (most probably) consisted of some inflammable petroleum compound, potassium nitrate to supply oxygen, and quick lime to supply further heat on reaction with water. It burned vigorously upon contact with water and hence could be used to destroy ships, which were made mainly of wood (Fig. 6.6).

The Greeks of the Byzantine Empire, for example, used this in 673 AD to repel the Arab naval onslaught on Constantinople. It is quite possible that, but for this 'surprise' weapon, the Arabs would have taken Constantinople; and world history would have been radically different [7].

The Arab scholars translated texts from Persia and India into Arabic, along with Ptolemy's *Almagest* (second century) and works from Euclid and Archimedes. These works helped further the growth of Astronomy. One of the Arab astronomers to study Ptolemy in detail was Albategnius (858–929 AD) also known as Al-Battani, who improved Ptolemy's work in several respects. He noticed, for instance, that the

Fig. 6.6: Depiction of the ‘Greek Fire’ in the Madrid Skylitzes manuscript [16]. This concoction was possibly invented by the seventh century Greek alchemist, Callinicus, to fight the Arabs. It contained essentially three ingredients: an inflammable petroleum compound, potassium nitrate, and quick lime. This mixture would burn vigorously on coming into contact with water and hence could be used for destroying ships (which were essentially made of wood). The Greeks of the Byzantine Empire used this in 673 AD to repel the Arab naval onslaught on Constantinople. It is believed that this ‘surprise’ weapon played a crucial role in that battle.

location at which the Sun appeared the smallest — when the Earth is farthest from the Sun — is not fixed but changes in a periodic manner. Al-Battani also obtained more accurate values for the length of the year, the precise time of the equinox, and the inclination of the Earth’s axis to the plane of revolution. It was his value for the length of the year that was used in the Gregorian reform of the Julian calendar seven centuries later (see Chap. 21).

Pointing out some of the errors that the earlier scholar had made with mapping planetary motion, Al-Battani also included a comprehensive star catalogue; solar, lunar, and planetary tables; and trigonometric tables and diagrams. Al-Battani’s findings and books strongly influenced later astronomers, like Kepler (1571–

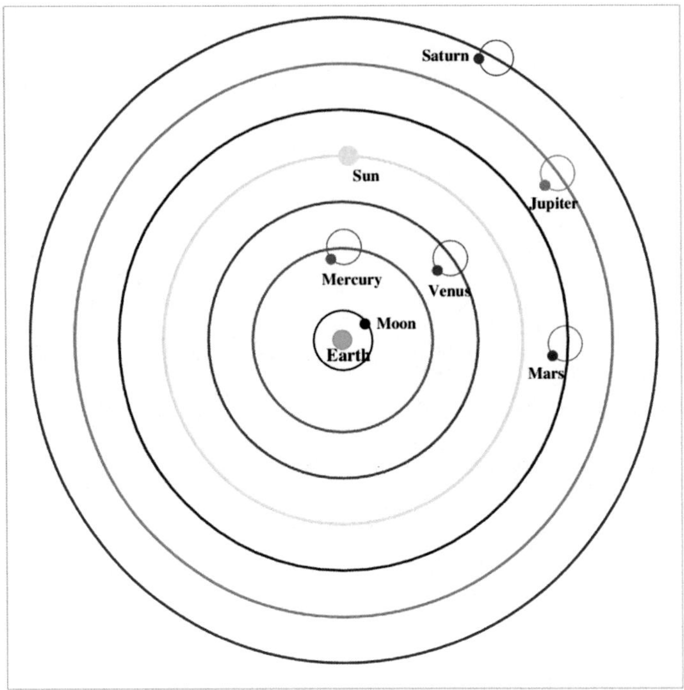

Fig. 6.7: One version of Ptolemy's geocentric universe has the Earth at the centre and the planets, Moon, and Sun orbiting around it. Ptolemy's *Almagest* was one of the many works translated into Arabic by the Arab scholars. Some of them, notably Albategnius (858–929 AD), also known as Al-Battani, corrected several errors in it in his commentary. In spite of the doubts expressed by the Arab scholars, *Almagest* — and the geocentric model described in it — continued to exert a major influence in Europe later on.

1630 AD), Galileo (1564–1642 AD), and Tycho Brahe (1546–1601 AD). While pointing out that Ptolemy's ideas of planetary motion (Fig. 6.7) were wrong, Al-Battani believed that the actual principles governing their motion were yet to be discovered. This refutation of Ptolemaic theory became a recurring theme over the centuries of Islamic scholarship, though it never managed to produce the heliocentric universe. He made significant contributions in mathematics and in particular trigonometry, and used trigonometrical methods for his calculations (rather than geometry, as Ptolemy had done). These advances in mathematics were used not only in astronomy, but also for many applications in engineering.

Box 6.2: Does the Earth Rotate?

The major difference between heliocentric and geocentric models lay in their appreciation of whether the Earth went around the Sun once a year or not. But what about the rotation of the Earth about its axis? Surely even the very ancient civilisations would have noticed that the Sun and the stars, say, rise in the east and set in the west each day; how does that come about?

Many members of the Pythagorean school, the earliest probably being Philolaus (470–385 BC), did indeed believe that the Earth rotated about an axis causing the daily rising and setting of heavenly bodies. However, Aristotle criticized the ideas of Philolaus and advocated a heavenly sphere of fixed stars which rotated about the Earth. This was accepted by later astronomers — in particular, Claudius Ptolemy — who thought that the Earth would feel continuous gale force winds if it rotated. This idea held sway for centuries in Europe.

Aryabhata (476–550 AD), a leading astronomer from India, had a clear notion of the Earth rotating around its axis, thereby causing the apparent daily motion of the heavens. In his influential textbook *Aryabhateya*, written around 500 AD, there is a verse which explicitly states: "Just as a man in a boat going in one direction sees the stationary things on the bank as moving in the opposite direction, in the same way to a man at Lanka the fixed stars appear to be going westward."

Arab scholars might have had access to this textbook, though historians are not sure about this. In any case, Arab scholars did acquire this knowledge, viz., that the Earth rotates about its axis, around the tenth century, in notable disagreement with Ptolemy. The astronomer, Al-Sijzi (945–1020 AD) [with the full name, Abu Sa'id Ahmed ibn Mohammed ibn Abd al-Jalil al-Sijzi!], in fact designed astronomical instruments based on the fact that "the motion we see is due to the Earth's movement and not that of the sky" [8].

It took several more centuries before this idea caught on in Europe. Copernicus (1473–1543) was probably the first person to clearly understand and state that the Earth rotated about its axis once every day. He did not accept the objections attributed to Ptolemy and Aristotle — viz., that such a rotation would cause gale force winds — because he felt that the movement of the stars could cause even more violent effects. In this aspect, Copernicus also acknowledged the contribution of the Pythagoreans.

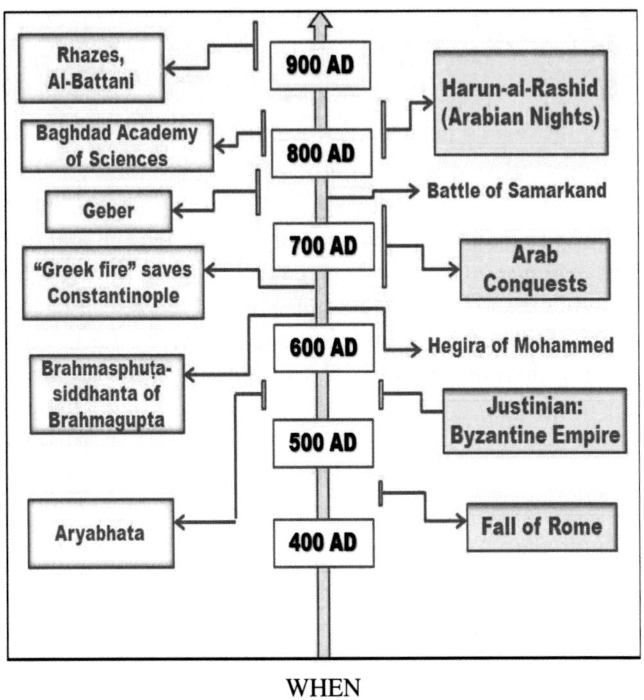

WHEN

Notes, References, and Credits

Notes and References

1. See, e.g.,
   Afnan, Soheil M. (1958), *Avicenna: His Life and Works*, G. Allen and Unwin, London [OCLC 31478971].
   Goodman, Lenn E. (2006), *Avicenna*, Cornell University Press, New York [ISBN 0-415-01929-X].
   Langermann, Y. T. (Ed.) (2010), *Avicenna and his Legacy. A Golden Age of Science and Philosophy*, Brepols Publishers, Belgium [ISBN 978-2-503-52753-6].
2. More on Rhazes can be found in:
   Iskandar, Albert et al (1997), "Al-Razi" in *Encyclopedia of the history of science, technology and medicine in non-Western cultures* (2nd edn.), Springer, The Netherlands [ISBN 978-1402045592].
   Browne, Edward Granville (2001), *Islamic Medicine*, Goodword Books Pvt. Ltd., New Delhi [ISBN 81-87570-19-9].

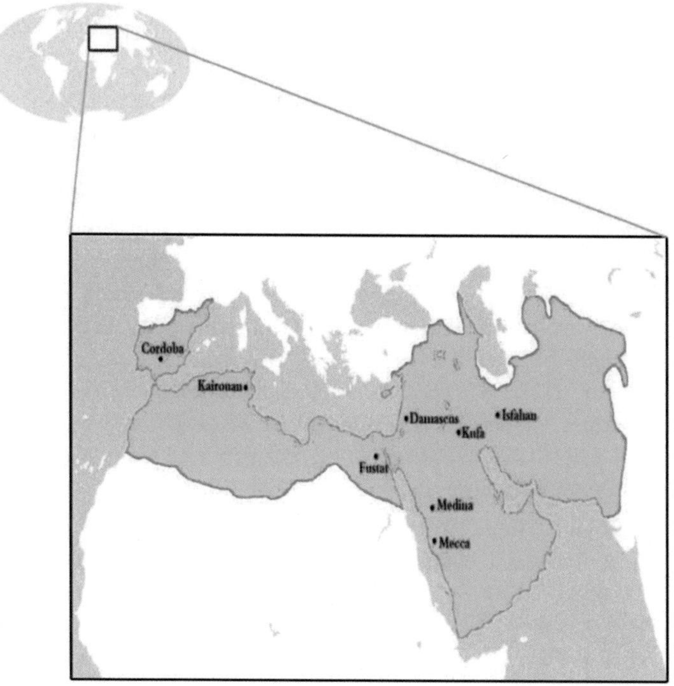

WHERE

3. The Egyptian scholar Youssef Ziedan started a project of collecting and examining the extant manuscripts of this work that are catalogued in many libraries around the world, including the Cambridge University Library and the Bodleian Library in the UK, and the Lane Medical Library at Stanford University. So far, 28 volumes of the encyclopedia have been published by him.
4. See, e.g.,

Tus, V. Minorsky (2000), in *The Encyclopedia of Islam – Vol. X* (Eds. P.J. Bearman, T. Bianquis, C.E. Bosworth, E. van Donzel, and W.P. Heinrichs), Brill Academic Publishers, Groningen.
Principe, Lawrence (2013),*The Secrets of Alchemy*, University of Chicago, Chicago [ISBN 0226682951].
Holmyard, E. J. (1931), *Makers of Chemistry*, Clarendon Press, Oxford.
Chisholm, Hugh (1910), "Geber" in *Encyclopedia Britannica* (11th edn.), pp. 545–546.

5. Needham, Joseph and Tsien, Tsuen-Hsuin (1985), *Science and Civilization in China: Volume 5: Chemistry and Chemical Technology, Part 1: Paper and Printing*, Cambridge University Press, Cambridge [ISBN 0-521-08690-6].
6. For the history of gunpowder and related topics, see, e.g.,
   Chase, Kenneth Warren (2003), *Firearms: A Global History to 1700*, Cambridge University Press, New York [ISBN 978-0-521-82274-9].
   Needham, Joseph (1987), *Science and Civilization in China: Military technology: The Gunpowder* (Volume 5, Part 7) Cambridge University Press, New York [ISBN 978-0-521-30358-3].
   Andrade, Tonio (2016), *The Gunpowder Age: China, Military Innovation, and the Rise of the West in World History*, Princeton University Press, Princeton [ISBN 978-0-691-13597-7].
   Hobson, John M. (2004), *The Eastern Origins of Western Civilization*, Cambridge University Press, Cambridge [ISBN 978-0521547246].
   Partington, J. R. (1999), *A History of Greek Fire and Gunpowder*, Johns Hopkins University Press, Baltimore [ISBN 0-8018-5954-9].
7. See, e.g., Roland, Alex (1992), *Secrecy, Technology, and War: Greek Fire and the Defense of Byzantium*, Technology and Culture **33** (4), 655–679 [doi:10.2307/3106585, JSTOR 3106585].
8. See:
   Young, M. J. L., (Ed.) (2006), *Religion, Learning and Science in the Abbasid Period*, Cambridge University Press, p. 413, Cambridge, UK [ISBN 9780521028875].
   Seyyed Hossein Nasr (1993), *An Introduction to Islamic Cosmological Doctrines*, pp. 135–136, State University of New York Press, New York [ISBN 0-7914-1516-3].

Figure Credits

9. Figure 6.1 (left) courtesy: https://commons.wikimedia.org/wiki/File:Avicenna-miniatur.jpg (from public domain).
10. Figure 6.1 (right) courtesy: Photo: Wolfgang Volk, Berlin. Reproduced here with permission.
11. Figure 6.2 courtesy: Gerardus Cremonensis /Wikimedia Commons / Public Domain; https://commons.wikimedia.org/wiki/File:Al-RaziInGerardusCremonensis1250.JPG (from public domain).
12. Figure 6.3 (left) courtesy: Muhammad ibn Zakariya al-Razi / Wikimedia Commons / Public Domain.
    https://commons.wikimedia.org/wiki/File:Colof%C3%B3n-Libro_de_Medicina_de_Razi.jpg (from public domain).

13. Figure 6.3 (right) courtesy: https://commons.wikimedia.org/wiki/File:Qanun_(Fil-Tibb),_Canon_(of_Medicine),_vol._5,_by_Ibn_Sina_(Avicenna),_Iran_or_Iraq,_dated_444_AH,_1052_AD,_watercolor_and_ink_on_paper_-_Aga_Khan_Museum_-_Toronto,_Canada_-_DSC06338.jpg (from public domain).
14. Figure 6.4 courtesy: https://commons.wikimedia.org/wiki/File:Making_Paper.gif (from public domain).
15. Figure 6.5 courtesy: Pericles of Athens / Wikimedia Commons/Public Domain. https://commons.wikimedia.org/wiki/File:Chinese_Gunpowder_Formula.JPG (from public domain).
16. Figure 6.6 courtesy: Anonymous / Wikimedia Commons/Public Domain. https://commons.wikimedia.org/w/index.php?title=File:Greekfire-madridskylitzes1.jpg (from public domain).

7 The Indo-Arabic Numerals

In 773 AD, at the height of Arab splendour, a man from distant India appeared at the court of the Caliph, Al-Mansur, in Baghdad. This traveller brought with him several volumes of writings from India. Al-Mansur had the sense to get them promptly translated into Arabic and, in a few decades, several Arab scholars assimilated their contents.

One among these scholars was [1] Abu Jafar Mohammed ibn Musa al-Khwarizmi (780–850 AD). His full name, when translated, means 'Mohammed, the father of Jafar and the son of Musa, the Khwarizmian', the last word originating from the Persian province of Khoresem. This man was one of the greatest mathematicians of the Arab world and he realized at once the importance of the number system used in the Indian writings. In fact, he wrote a small book explaining various aspects of the use of these numerals around 820 AD.

The original of this book is lost but there is a fair amount of evidence to suggest that it reached Spain around 1100 AD; there it was translated into Latin by an Englishman, Robert of Chester. This translation is probably the earliest known introduction of Indian numerals to the West. The manuscript begins with the words, '*Dixit Algoritmi: laudes deo rectori nostro atque defensori dicamus dignas*' ('Algoritmi has spoken; praise be to God, our Lord and our Defender', with the Arab name Al-Khwarizmi having been transliterated into Algoritmi in Latin). In later years, careless readers of the book started attributing the calculational procedures described in the book to a person called Algoritmi; that is how we got the term 'algorithm' for any computational procedure.

T. Padmanabhan and V. Padmanabhan, *The Dawn of Science*,
https://doi.org/10.1007/978-3-030-17509-2_7

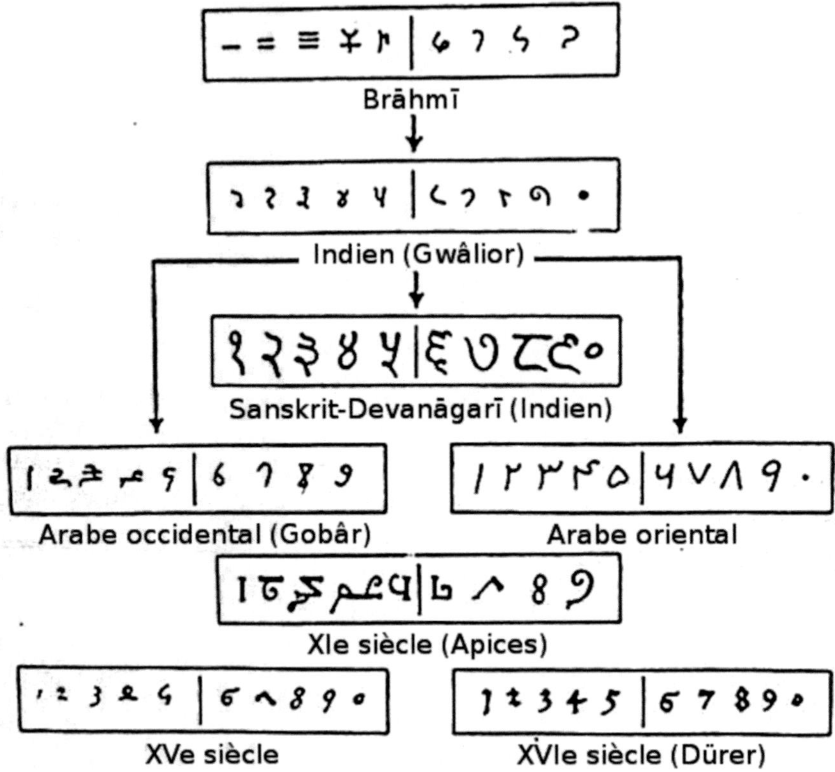

Fig. 7.1: The genealogy of modern numerals, starting from India, through the Arabs to the West [5]. Several volumes of scholarly writings from India reached the court of Caliph, Al-Mansur, in Baghdad in 773 AD. The Caliph arranged for them to be translated into Arabic and amongst several Arab scholars who assimilated their contents was [1] Al-Khwarizmi, one of the foremost mathematicians of the Arab world. Having understood the importance of the number system described in the Indian works, he wrote a small book (around 820 AD), explaining several aspects of its use. Around 1100 AD, this book reached Spain; there, it was translated into Latin by an Englishman, Robert of Chester. This was probably the earliest introduction of Indian numerals to the West.

The use of Indian numerals was picked up by several scholars and was taught in major cities, thereby leading to the use of zero becoming well established (Fig. 7.1). In fact, Al-Khwarizmi himself says explicitly: "When nothing remains . . . put down a small circle so that the place be not empty ... and the number of places is not diminished and one number is not mistaken for the other."

However, the new system was not easily accepted by the average man in the street; ultimately, what tilted the balance in its favour was not the scholarly exposition, but purely commercial considerations! By the end of the first millennium, Italy had grown to be a major mercantile power around the Mediterranean. Italian ships were used for the Crusades, Italian bankers were major money-lenders, and Venice, Genoa, and Pisa rose as cities of prominence. The traders and merchants were quick to realize the advantages of the Indo-Arabic number system [2]. Blessed by big business, the system was there to stay — a familiar pattern which occurred again and again through history! For example, the *Margarita Philosophica* (the philosophic pearl), a beautifully illustrated encyclopedia (Fig. 7.2), authored by the monk Gregor Reisch (1467–1525 AD), which was widely used as a university textbook in the early sixteenth century, discusses arithmetic using Indo-Arabic numerals compared to the use of a counting board.

Al-Khwarizmi also wrote another influential book called *Al-jabr-wa'l Muquabala*, (which could be translated as 'The science of transposition and cancellation'). In this work, he gives a detailed exposition of the fundamentals of the subject which has come to be called 'algebra'. Al-Khwarizmi discusses in a systematic manner — among other things — the solutions of linear and quadratic algebraic equations. The clarity of the discussion in this book has made later workers call Al-Khwarizmi, quite justifiably, the 'father of algebra'.

The key aspect of algebra, in contrast to, say, arithmetic, is the use of symbols to represent mathematical constructs and operations. Indeed, this has also evolved over ages. The earliest discussions in algebra, both in the East and the West, were rather rhetorical. Questions and answers were given in the form of dialogues and no symbols were used. Two of the earliest mathematicians to realize the powers of symbolic manipulations were Diophantus ($\sim$ 201–285 AD) in Greece and Brahmagupta (598–668 AD) in India [3]. Diophantus used symbols to denote unknown quantities, various powers of an unknown quantity, reciprocals, and equality. He also used Greek letters to denote different numerals. The system followed by Brahmagupta was somewhat more elaborate. Addition was indicated by just placing the terms next to each other, subtraction by placing a dot over the term to be deducted, multiplication by writing the Sanskrit letter *bha* (the first letter of *bhavitha*, the product), and the square root by the prefix *ka* (from the word *karana*). The first unknown variable in the problem was denoted by *ya* and additional unknowns were indicated by the initial syllables of various colours.

Many familiar mathematical symbols we use today came into existence gradually over the centuries. The 'equals' sign (=) was due to Robert Recorde (1512–1558 AD), appearing (Fig. 7.3) in his *The Whetstone of Witte* (1557) [4]. The 'plus' and

Fig. 7.2: The title page [6] of Gregor Reisch's book *Margarita Philosophica* (1503). Reisch pursued efforts to build encyclopedic works that would gather together all available knowledge. This work contained 12 books on the arts and sciences and served as a textbook. The seven 'liberal arts' are shown around the three-headed figure in the center. (These are the seven topics commonly taught at universities: the trivium of logic, rhetoric, and grammar and the quadrivium of arithmetic, music, geometry, and astronomy.) Arithmetica is seated with a counting board in the middle.

The Arte

as their woorkes doe extende) to distincte it onely into twoo partes. Whereof the firste is, *when one nomber is equalle unto one other.* And the seconde is, *when one nomber is compared as equalle unto .2. other nombers.*

Alwaies willyng you to remēber, that you reduce your nombers, to their leaste denominations, and smalleste formes, before you procede any farther.

And again, if your *equation* be soche, that the greateste denomination *Cosike*, be ioined to any parte of a compounde nomber, you shall tourne it so, that the nomber of the greateste signe alone, maie stande as equalle to the reste.

And this is all that neadeth to be taughte, concernyng this woorke.

Howbeit, for easie alteratiō of *equations*. I will propounde a fewe exāples, bicause the extraction of their rootes, maie the more aptly bee wroughte. And to auoide the tediouse repetition of these woordes: is equalle to: I will sette as I doe often in woorke use, a paire of paralleles, or Gemowe lines of one lengthe, thus: =====, bicause noe .2. thynges, can be moare equalle. And now marke these nombers.

1. 14.ze.—+—.15.ϙ=====71.ϙ.
2. 20.ze.———.18.ϙ=====.102.ϙ.
3. 26.ʒ.—+—10ze=====9.ʒ—10ze—+—213.ϙ.
4. 19.ze—+—192.ϙ=====10ʒ—+—108ϙ—19ze
5. 18.ze—+—24.ϙ.=====8.ʒ.—+—2.ze.
6. 34ʒ———12ze=====40ze—+—480ϙ—9.ʒ.

1. In the firste there appeareth. 2. nombers, that is 14.ze.

Fig. 7.3: The familiar mathematical symbols we use today took centuries to come into existence. The two earliest mathematicians to appreciate and use the powers of symbolic manipulations were Diophantus in Greece and Brahmagupta in India. The figure [7] shows the first occurrence of the 'equals' sign (=), which was due to Robert Recorde [4] and appeared in his *The Whetstone of Witte* (1557). The 'plus' and 'minus' signs first appeared in print in a textbook on arithmetic written by Johannes Widmann (published in 1489), and the signs for multiplication and proportion were introduced much later by William Oughtred.

'minus' signs first appeared in print in an arithmetic textbook written by Johannes Widmann (1460–∼1498 AD) and published in 1489. The signs for multiplication and proportion were introduced by William Oughtred (1574–1660 AD). It was

Descartes (1596–1650) who brought in the present compact notation with indices a, a^2, a^3, etc. Finally, the notation π for the ratio between the circumference and the diameter of a circle was first used by the English writer, William Jones (1675–1749 AD), in 1706.

Box 7.1: Trigonometry and a Cavity in Your Nose

In trigonometry, we associate with each angle certain ratios usually called sine, tangent, and secant. (Three more ratios, cosine, cotangent, and cosecant, arise as complements of these three ratios.) Of these three, the terms 'tangent' and 'secant' have clear geometrical meanings and correspond to the standard definitions in Euclidean geometry (Fig. 7.4). How did the word 'sine' originate?

There is an interesting story behind this word 'sine'. Surprisingly enough, it came from a Sanskrit term jya-ardha ('half of a chord')! Of course, 'jya-ardha' does not sound anything like 'sine' and the transformation again involves the Arab legacy.

The Indian mathematician, Aryabhata (476–550 AD), used the term *jya-ardha* to denote the ratio that we now call sine, which, of course, makes complete geometrical sense. The ratio in question is indeed the length of the half-chord in a unit circle (Fig. 7.4). This term, abbreviated as *jya*, was converted phonetically to *jiba* by the Arabs. Following a standard practice in the Arabic language of dropping the vowels in the written version, this was truncated to just *jb*. The term *jiba* has no meaning in Arabic except in this specific technical context.

Later writers, who came across *jb* as a shortened version for *jiba* (which, of course, appeared meaningless to them), decided to 'correct' it to *jaib* — by adding wrong vowels at the wrong places! — which is an Arabic word meaning 'cove' or 'bay'. Later on, an Italian translator, Gerard of Cremona (1114–1187 AD), while translating technical terms from Arabic to Latin, translated *jaib* literally to the Latin equivalent *sinus*. This became 'sine' in English. That is how a cavity in the upper nose and a trigonometric ratio ended up having the same linguistic root!

The power of Arabic civilization to synthesize knowledge influenced several aspects of trigonometry. This subject, which was well developed in both India and Greece due to the stimulus given by astronomical observations, attained a

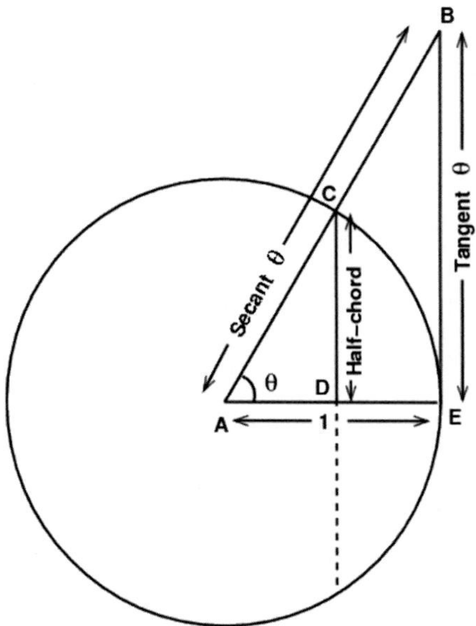

Fig. 7.4: Geometrical meaning of the trigonometric ratios. In a circle of radius 1 unit, draw a tangent BE and secant BA from the external point B. The lengths of BE and AB give, respectively, the value of 'tangent θ' (or $\tan\theta$) and 'secant θ' (or $\sec\theta$). Similarly, the length of the half-chord ('jya-ardha', in Sanskrit) CD is equal to 'sine θ'. (The way the Sanskrit word 'jya-ardha' got transformed into 'sine' is described in Box 7.1!)

unified form in the hands of the Arabs. In Greece, it was developed by Aristarchus ($\sim$ 310–230 BC), Hipparchus (around 140 BC), and most notably Claudius Ptolemy ($\sim$ 100–170 AD) in later years. In particular, Ptolemy constructed what he called a 'table of chords' which is equivalent — in modern language — to the trigonometric table for the sine of an angle. He did this using a very elegant geometrical procedure for all the angles at half-degree intervals. This work, of course, was developed further by the Arab scholars. Abu al-Wafa (940–998 AD), for example, produced the tables for sines and tangents at intervals of a quarter of a degree and these were used extensively by later scholars. Similar tables were constructed in India by the scholar and leading astronomer, Aryabhata (476–550 AD). As we saw in Chap. 6, he had also stated clearly, in his influential textbook *Aryabhateya*, that the Earth rotated about its axis. Arabic scholars acquired this knowledge around the tenth century, but it was only clearly established in Europe after the work of Copernicus (1473–1543).

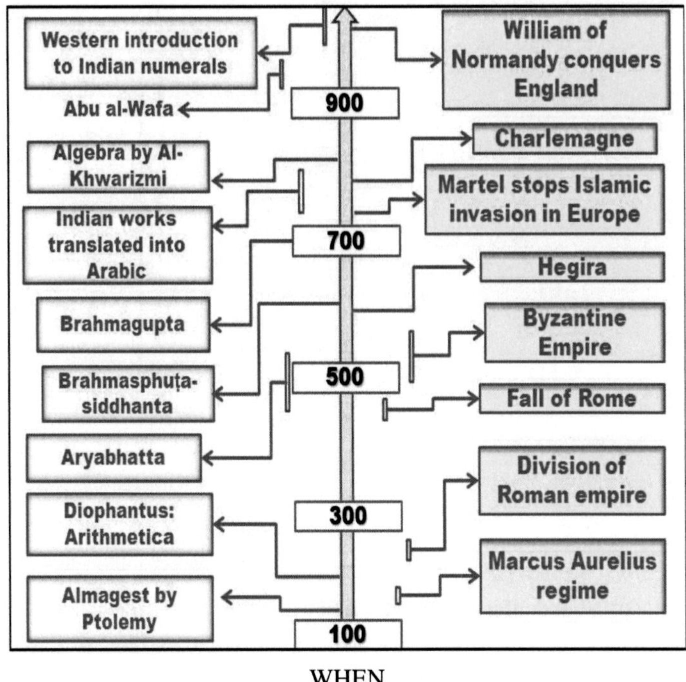

WHEN

Notes, References, and Credits

Notes and References

1. For more on Muhammad ibn Musa al-Khwarizmi, see, e.g.,
   Brentjes, Sonja (2007), "Khwarizmi: Muhammad ibn Musa al-Khwarizmi" in Thomas Hockey et al.(eds.), *The Biographical Encyclopedia of Astronomers*, Springer Reference, Springer, New York [ISBN 978-0-387-35133-9].
   Rosen, Fredrick (1831), The *Algebra of Mohammed Ben Musa*, Kessinger Publishing, US [ISBN 1-4179-4914-7].
   Berggren, J. Lennart (1986), *Episodes in the Mathematics of Medieval Islam*, Springer Science+Business Media, New York [ISBN 0-387-96318-9].
   Boyer, Carl B. (1991), "The Arabic Hegemony" in *History of Mathematics* (Second ed.), John Wiley and Sons, Inc., New York [ISBN 0-471-54397-7].
2. For more on the developments of Indo-Arabic number system, see:
   Ifrah, Georges (1998), *The Universal History of Numbers: From Prehistory to the Invention of the Computer*, Harvill, London [ISBN 978-1-860-46324-2].

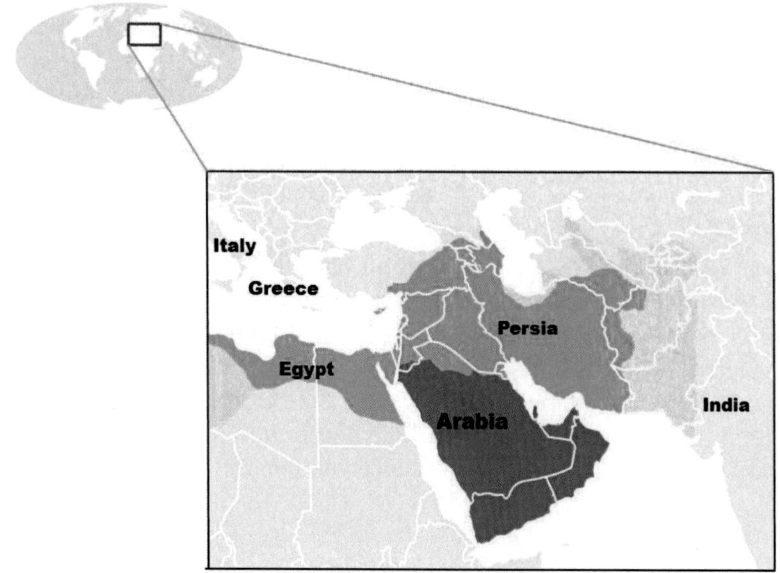

WHERE

Menninger, Karl (2013), *Number Words and Number Symbols: A Cultural History of Numbers* (translated by Paul Broneer), Courier Corporation, US [ISBN 978-0-486-31977-3].

Plofker, Kim (2009), *Mathematics in India*, Princeton University Press, Princeton [ISBN 978-0-691-12067-6].

3. More on mathematical symbolism can be found in:

Boyer, C.B. (1991), *A history of mathematics*, Wiley, New Jersey [ISBN 978-0471543978].

Kline, M. (1990), *Mathematical thought from ancient to modern times*, Oxford University Press, Oxford [ISBN 978-0195061352].

Atkins, Peter (2004), *Galileo's Finger: The Ten Great Ideas of Science*, Oxford University Press, Oxford [ISBN 9780191622502].

4. *The Whetstone of Witte* is the shortened title of Robert Recorde's mathematics book published in 1557. The full title is a mouthful: "The whetstone of witte, whiche is the seconde parte of Arithmetike: containyng thextraction of Rootes: The Cubike practise, with the rule of Equation: and the woorkes of Surde Nombers." The topics in the book

include whole numbers and the extraction of roots and irrational numbers. The work is notable for containing the first recorded use of the equals sign and also for being the first book in English to use the plus and minus signs.

Figure Credits

5. Figure 7.1 courtesy: Piero/ Wikimedia Commons /Public Domain. https://commons.wikimedia.org/wiki/File:Numeration-brahmi_fr.png (from public domain).
6. Figure 7.2 courtesy: Wellcome Library, London, Title: L0025814 G. Reisch, Margarita philosophica Photo number: L0025814. https://commons.wikimedia.org/wiki/File:G._Reisch,_Margarita_philosophica_Wellcome_L0025814.jpg (published under CC by 4.0).
7. Figure 7.3 courtesy: Rnm at Polish Wikipedia / Wikimedia Commons/Public Domain. https://commons.wikimedia.org/wiki/File:Robert_Recorde-The_Whetstone_of_Witte_1557.gif (from public domain).

8

The Printing of a Page

The first four centuries of the previous millennium, the period 1000–1400 AD, witnessed significant instability and turmoil in much of Asia and Europe. The Arabs started fighting among themselves and their empire started to break up. In Europe, several smaller states sprang up from the remnants of the Western Roman Empire, but they, too, were never at peace with each other; in addition, they were bent on carrying on the 'holy' Crusades against the Turks — more often than not — for political ends. This was also the period (about 1220–1250) which saw the rise of the Mongols in Asia under Genghis Khan ($\sim$ 1162–1227) and his descendants, who destroyed the cities of Bokhara, Samarkand, and Baghdad. To cap it all, there was an outbreak of the 'Great Plague' (around 1348) which spread through Europe, North Africa, Russia, and even parts of China, wiping out entire populations.

Clearly, this was not an atmosphere in which science could prosper, and indeed it suffered. However, at the turn of the fifteenth century, there came an invention which totally transformed the history of mankind. It was the invention, made some time around the year 1430, of the so called 'movable printing process' by Johannes Gutenberg (1400–1468). This single invention exerted more influence on the Scientific Revolution than all the scholarly expositions of several medieval scientists put together!

In spite of its importance, the story of the printing press is a tragic one, as far as the inventor himself is concerned. Gutenberg was the son of a patrician in the city of Mainz (in Germany). There he was associated with the goldsmith's guild and learnt several skills in working with metals. Unfortunately, his colleagues, who

T. Padmanabhan and V. Padmanabhan, *The Dawn of Science*,
https://doi.org/10.1007/978-3-030-17509-2_8

became envious of his growing skills and prosperity, managed to get him exiled from Mainz.

Texts used to be reproduced mechanically long before Gutenberg's time, but it was done in a really primitive form. All the ancient civilizations used some form of seals and insignia, which constituted the earliest form of printing. As early as the second century AD, the Chinese had the three essential elements required for printing: paper, the manufacture of which was known to them; ink, the basic formula of which had been known for centuries; and surfaces, on which letters and texts could be conveniently engraved.

In fact, books had indeed been printed as early as 764 AD in Japan and China using rather primitive techniques – the text was cut out of wood and, by spreading ink on it, the woodcut was transferred to paper. The earliest example we have is the Diamond Sutra [1], produced in 868 AD. A copy of the Chinese version of the Diamond Sutra, possibly from the time of the Tang dynasty, was found among the Dunhuang manuscripts by the Buddhist monk Wang Yuanlu in 1900. (Incidentally, this book is also the first creative work with an explicit public domain licence, since it states that it was created "for universal free distribution"!) But this process of wood-block printing was so laborious that it never caught on, either in the Arab world or — later on — in Western Europe.

Instead, in Gutenberg's time, books in Europe were essentially reproduced by copying them laboriously by hand. This meant that books were few (and expensive) and only monasteries, universities, and very rich people could possess them. What is more, every copy introduced some chance for errors to creep in, which was unthinkable in the case of religious texts. (In fact, Jewish copyists of the Bible took the elaborate precaution of counting the total number of letters whenever they finished copying something!)

Box 8.1: Just for the Record

The earliest records we have are small, marked clay tokens that represented ancient account keeping — like a number of sheep, bags of grain, etc. These records date back to 9000 BC in prehistoric Mesopotamia. Moving ahead, around 3000 BC we have evidence of the Sumerians writing in the form of pictograms made of pictures related to the words or their sounds. Egyptian hieroglyphs (meaning 'sacred carving' in Greek), with over 700 different signs, belong to the same category. Later on, the Sumerians developed cuneiform writing ('wedge writing', in Latin), using a wedge-shaped stylus to write on damp, clay tablets. These tablets were then baked to preserve the writing. The Sumerians produced a large number of clay tablets describing their laws, transactions, and literature.

While the Egyptians also wrote on the same kind of surfaces as the Sumerians (clay, stone, bone, metal, etc.), their favourite technique was to use reed pens on papyrus scrolls. The papyrus plant, which is a tall reed that grows around the Nile River, made this rather easy. The longest Egyptian scroll archaeologists have found is over 133 feet long, just 18 feet shorter than the Statue of Liberty!

Papyrus continued to be the most popular writing medium for centuries, and both the Greeks and the Romans used it. Sometime during the third century, the Romans began to sew together the folded papyrus sheets and put them between two wooden covers — and the first book was born. The wooden covers led to the name *codex*, which initially meant 'a piece of wood' in Latin.

Parchments, made of animal skin, had replaced papyrus as the main writing surface by the fourth century AD. There is an interesting legend associated with this switch, coming to us from the Roman writer Pliny the Elder (23–79 AD). Since most of the papyrus grew in the Nile region of Egypt, it gave Egyptians a monopoly over this resource. Apparently, King Ptolemy of Egypt was so jealous of the library of the King of Pergamum in Asia Minor, that he banned the export of papyrus to Pergamum. This embargo forced the people of Pergamum to use fine animal skins as an alternative writing medium. (The word 'parchment' comes from the Latin word 'pergamena', possibly named after the city of Pergamum.) Parchment had several advantages: first, one could use both sides of a parchment; second, it lasted longer; and third — most importantly — the ink on the parchment could be removed by scraping it off and the parchment could be reused. Such a reused parchment was a palimpsest. (We came across one in Chapter 4.)

In Asia, 'books' were made from bamboo, tree bark, and palm leaves. Thin sheets of palm leaves were cut into strips and a cord was threaded through holes drilled along the edge of the strips. The palm leaf pages were usually kept between two pieces of wood or bark for storage. Before paper became popular, the Chinese also wrote on silk scrolls, thin bamboo, and wood. After the fifth century, however, paper scrolls were used almost exclusively. Paper spread from China through the Arab world to Europe by the fourteenth century, and by the late fifteenth century it had replaced parchment for all practical purposes.

Obviously, the copying of such texts, undertaken in the medieval days, was a time-consuming task. The text, after copying, would be checked for copying errors, corrected, and then the 'rubricator' would add to the chapters the titles, initial capitals, and paragraph marks. (The word 'rubric' comes from the Latin word for red, because the headings were often in red.) Given the time and effort involved, the books could only be preserved in libraries (in the ancient days) and in monasteries (in medieval Europe) and it was impossible for ordinary individuals to own books. This meant there was a risk of mass destruction of knowledge in times of crisis. For example, the library of Alexandria was totally destroyed during political skirmishes in the fifth century. Similarly, the great library of Constantinople was destroyed when the city was first sacked by the Crusaders in 1204 AD, and by the Turks in 1453 AD.

The invention of printing changed the situation completely, so that such disasters could never happen again. It was now possible to produce hundreds of copies of a book and the common man could get access to them. The spread of knowledge caused by this single invention was enormous.

Gutenberg's genius was in realizing that, by producing a set of small and durable metal seals, each representing just a single letter, printing could be made much more efficient. The letters could be assembled to form a page (Fig. 8.1); and once the page was printed, the letters could be disbanded and reassembled to make up the next page. So, here was a possibility of printing unlimited copies of a book using the same basic set of printer's seals.

As often happens, though the idea was simple, its execution was not [2]. To succeed, Gutenberg needed to develop techniques that would allow him to form tiny letters out of metal pieces and, importantly, of uniform quality. It was also necessary to produce good quality ink. All this took Gutenberg almost twenty years.

During these years, however, Gutenberg also got involved in several lawsuits with his partners and financiers and — unfortunately — lost most of them. Gutenberg was naturally anxious to keep his project a secret to prevent others from cashing in quickly on his idea. But the legal proceedings brought the nature of the project he was working on out into the open, and many of his financiers were quick to pounce on the possible profits from his idea. Notable among them was Johann

Fig. 8.1: Reproduction of a Gutenberg-era press on display at the Printing History Museum in Lyon, France [3].

Fust (1400–1466) who won a crucial lawsuit against Gutenberg around 1450. As a result of this court verdict — in which the court decided that Gutenberg had no reasonable means of paying back his creditors — Gutenberg had to hand over his entire printing press and the tools to Fust.

Ironically, this was just around the time when Gutenberg was getting ready for the printing of a beautiful version of the Bible — which is well known today as the Gutenberg Bible — printed in double columns with 42 lines in Latin on each page (Fig. 8.2). It is widely considered to be an exquisite work of art and remains one of the most expensive books in the world today.

Though Gutenberg died in debt, a broken man, his invention *did* transform the world. The technique of printing spread quickly all over Europe. By 1470 AD, printing presses were operating in Italy, Switzerland, and France. In 1476 AD, the first printing press in England was established by William Caxton. By 1535 AD the invention of printing had crossed the Atlantic and a press was established in Mexico City.

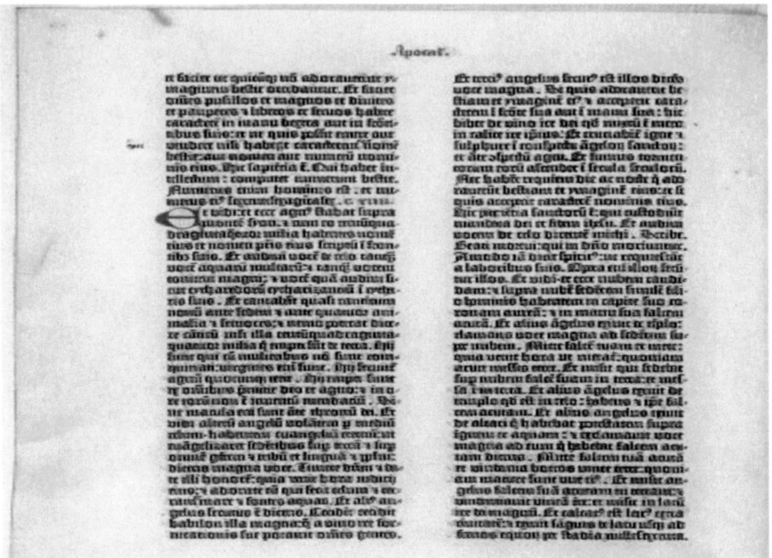

Fig. 8.2: A page from the Gutenberg Bible [4]. The Gutenberg Bible (also known as the 42-line Bible or the Mazarin Bible) was the first major book to be printed using movable metal type, and it marked the beginning of the age of the printed book in the West. Because of its aesthetic and artistic qualities, the book has acquired an iconic status. The 36-line Bible, believed to be the second printed version of the Bible, is also sometimes referred to as a Gutenberg Bible, but it may be the work of another printer.

Though initially only religious texts came out of the press, it wasn't long before scholars and various others started using this medium to spread their good word (see Fig. 8.3). It is, for example, very unlikely that Martin Luther (1483–1546 AD) would have succeeded in his rebellion against the Church but for the fact that he could print and distribute a large number of pamphlets. Since printing provided cheaper books, literacy went up and the ranks of the educated community swelled. This, in turn, heralded the coming of the Scientific Revolution.

A TREATISE ON THE ASTROLABE PROLOGUS

LITEL LOWIS MY SONE, I HAVE PERCEIVED wel by certeyne evidences thyn abilite to lerne sciences touchinge noumbres and proporciouns; & as wel considere I thy bisy preyere in special to lerne the Tretis of the Astrolabie. Than, for as mechel as a philosofre seith, he wrappeth him in his frend, that condescendeth to the rightful preyers of his frend, therfor have I geven thee a suffisaunt Astrolabie as for oure orizonte, compowned after the latitude of Oxenford; upon which, by mediacion of this litel tretis, I purpose to teche thee a certein nombre of conclusiouns apertening to the same instrument. I seye a certein of conclusiouns, for three causes. The furste cause is this: truste wel that alle the conclusiouns that han ben founde, or elles possibly mighten be founde in so noble an instrument as an Astrolabie, ben unknowe perfitly to any mortal man in this regioun, as I suppose. Another cause is this; that sothly, in any tretis of the Astrolabie that I have seyn, there ben some conclusiouns that wole nat in alle thinges performen hir bihestes; & some of hem ben to harde to thy tendre age of ten yeer to conseyve. This tretis, divided in fyve parties, wole I shewe thee under ful lighte rewles & naked wordes in English; for Latin ne canstow yit but smal, my lyte sone. But natheles, suffise to thee thise trewe conclusiouns in English, as wel as suffyseth to thise noble clerkes Grekes thise same conclusiouns in Greek, & to Arabiens in Arabik, and to Jewes in Ebrew, & to the Latin folk in Latin; whiche Latin folk han hem furst out of othere diverse

Fig. 8.3: A printed page from the medieval instruction manual on the astrolabe by Geoffrey Chaucer (1343–1400 AD), notable for being written in English prose and for describing a scientific instrument [5]. The treatise is considered by many to be the oldest work in English, written about a scientific instrument having an original date of 1392 AD. In those days, it was admired for its clarity in explaining difficult concepts. Chaucer's exact source is unknown, but most of his description goes back, directly or indirectly, to *Compositio et Operatio Astrolabii*, a Latin translation of Messahala's Arabic treatise of the eighth century.

Box 8.2: Gutenberg Bible

Set in Latin, the Gutenberg Bible is an edition of the Vulgate, printed in Mainz (Germany) during the 1450s. Technically speaking, of course, it was not Gutenberg who brought this book out, but Fust and his collaborators. In March 1455, the future Pope Pius II wrote that he had actually seen pages from the Gutenberg Bible which were being displayed in Frankfurt to promote the edition. It is not known how many copies were printed, with the letter citing sources for both 158 and 180 copies. Around 49 copies (or substantial portions of copies) have survived till today, and they are considered to be among the most valuable books in the world.

Each of the surviving copies have stories behind them. Bridwell Library of the Southern Methodist University, Texas, for example, has 31 leaves of the Gutenberg Bible. These came from an incomplete second volume discovered in 1828 at a farmhouse near Trier, Germany. The volume, which eventually became the property of a Jewish chemist, was sold in 1937 to raise money for his escape from Nazi Germany to London!

The buyer turned it over to Charles Scribner's Sons in New York in 1953. This firm removed 116 leaves of the New Testament (which are now at Indiana University) for a collector in Chicago and sold the 132 remaining leaves individually and in small groups. The largest of these groups, this 31-leaf fragment, went to John M. Crawford, Jr., of New York City, who sold it to Bridwell Library on 11 June 1970.

It is amusing to note that Gutenberg's invention of the printing press and its legacy forms an interesting contrast to the contributions of his more famous contemporary, Leonardo da Vinci (1452–1519 AD). Well-known for his artistic creations and often called the true Renaissance man, da Vinci had created — in his sketch books — very imaginative inventions. Unfortunately, none of them had the same practical impact as Gutenberg's much less glamorous creation! da Vinci's creations have come to us from *The Codex Atlanticus* (Atlantic Codex) — currently preserved at the Biblioteca Ambrosiana in Milan. This is a twelve-volume set of drawings and writings (in Italian), covering a great variety of subjects, including flying machines, weaponry, musical instruments, etc. His main fascination was with flight and he 'designed', again only in a sketch book, a flying machine, a helicopter, a parachute, and an instrument to measure wind speed. Unfortunately, many of these ideas were impractical and never seem to have matured into something that people could use in his day.

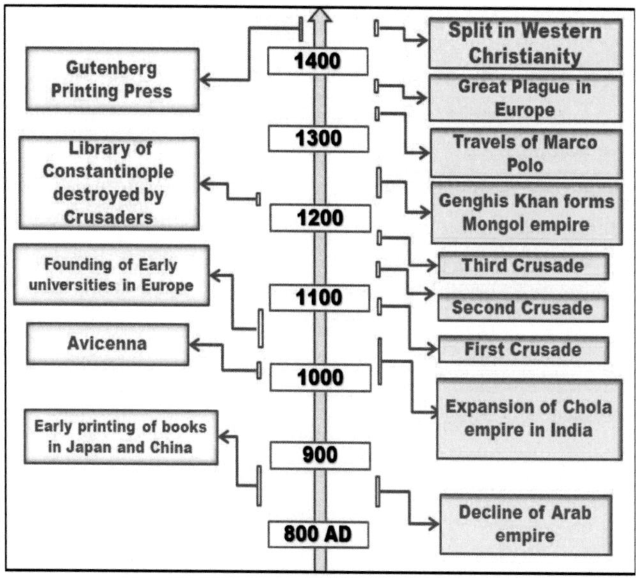

WHEN

Notes, References, and Credits

Notes and References

1. Wood, Frances and Barnard, Mark (2010), *The Diamond Sutra: The Story of the World's Earliest Dated Printed Book*, British Library, UK [ISBN-13: 978-0712350907].
2. For more on the story of printing and its impact, see, e.g.,
   Childress, Diana (2008), *Johannes Gutenberg and the Printing Press*, Twenty-First Century Books, Minneapolis [ISBN 978-0-7613-4024-9].
   Eisenstein, Elizabeth (1980), *The Printing Press as an Agent of Change*, Cambridge University Press, Cambridge [ISBN 0-521-29955-1].
   Eisenstein, Elizabeth (2005), *The Printing Revolution in Early Modern Europe* (2nd rev. edn.), Cambridge University Press, Cambridge [ISBN 0-521-60774-4].

Febvre, Lucien and Martin, Henri-Jean (1997), *The Coming of the Book: The Impact of Printing*, 1450–1800, Verso, London [ISBN 1-85984-108-2].
McLuhan, Marshall (1962), *The Gutenberg Galaxy: The Making of Typographic Man*, University of Toronto Press, Toronto [ISBN 978-0-8020-6041-9].

Figure Credits

3. Figure 8.1 courtesy: https://commons.wikimedia.org/wiki/File:Gutenberg.press.jpg (from public domain).
4. Figure 8.2 courtesy: Johannes Gutenberg / Wikimedia Commons /Public Domain. https://commons.wikimedia.org/wiki/File:Gutenberg_bible.jpg (from public domain).
5. Figure 8.3 courtesy: The works of Geoffery Chaucer /Wikimedia Commons/ https://commons.wikimedia.org/wiki/File:Chaucer_tratise_book.jpg (from public domain).

9

Exploring the 'Seven Seas'

By about 1260 AD, Kublai Khan (1215–1294) had set up a great Mongol empire in China. It was during this supremacy of the Mongol dynasty that China came into closer contact with Europe. With a fairly stable Mongol Empire stretching across the vast plains of Asia, it was easy for European travelers to visit China and establish regular trade links.

This trade and commerce brought rich dividends not just to the explorers and their kingdoms, but also to the science of geography, through the explorations of land and sea routes.

One amongst such traders who deserves a place in history [1] was Marco Polo (1254–1324), from Venice. His father and an uncle made a trip to Kublai Khan's China when Marco was still in his early teens. The Mongol leader was fascinated by the Venetian merchants and sent them back to Europe to bring Christian missionaries who would introduce Christianity to China. The Polos could not persuade the papacy to send the clergy, but Marco Polo accompanied his father and uncle on their second trip, reaching China in 1275. He had a cordial relationship with Kublai Khan and eventually became the Emperor's trusted diplomat. After spending nearly two decades in the Orient, Marco Polo finally returned to Venice in 1295. This was the first time Central Asia had been closely observed by Europeans.

Marco Polo, who held command of a Venetian fleet, was captured in a local naval battle in 1298 and had to spend a year in a Genovese prison. While in jail, he wrote his travelogue (rather pompously entitled *The Description of the World*), describing in detail the affairs of Asia and the Far East. The book was popular but

T. Padmanabhan and V. Padmanabhan, *The Dawn of Science*,
https://doi.org/10.1007/978-3-030-17509-2_9

Fig. 9.1: Marco Polo at Kublai Khan's court [10]. This miniature appears in an Italian book describing his travels, originally published during his lifetime and frequently reprinted and translated. Marco Polo, on his return from China after serving the Mongols for nearly two decades, wrote a detailed account of what he had seen in his book *The Description of the World*. It gave a lot of details about cities, canals, rivers, ports, and industry in China and neighboring areas.

its contents were not believed by most people. Given the routes he had followed and the places he had visited (Fig. 9.2), it is no wonder that the book was a bestseller! The main reason for disbelief was European prejudice, which did not accept the idea that there could be a great civilization and riches in the Far East. They even coined the term "Marco Milione" (Marco Millions) to describe the way Marco Polo dealt with large numbers in his descriptions of the East.

There was, however, one man who believed every word of what Marco Polo had to say. This was Christopher Columbus (1451–1506), another Italian explorer [2], who believed strongly that the almighty had chosen him to achieve great deeds. He wanted to acquire the riches of the Indies and Cathay (the names by which

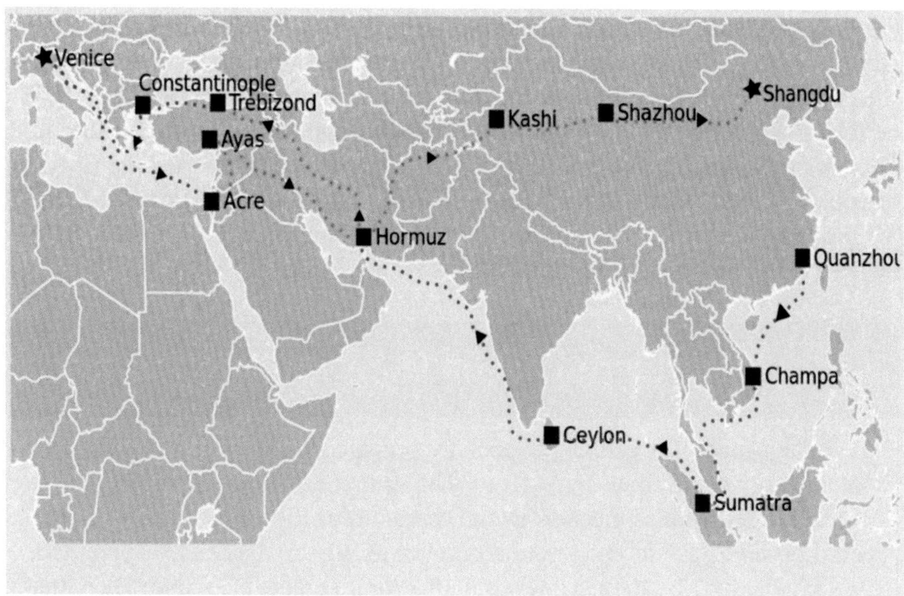

Fig. 9.2: Map of Marco Polo's voyages [11]. In 1271, Marco Polo left for the Mongol court with his father and uncle. After passing through Acre (now in Israel) they proceeded to Ayas (now in Turkey), crossed several fairly inhospitable deserts, and reached Hormuz on the Persian Gulf. The Polos then proceeded by land to the Mongol capital taking a series of traders' routes — which, later in the nineteenth century, became known as the Silk Route — reaching Shangdu in 1275. After spending nearly two decades as the trusted diplomat of Emperor Kublai Khan and travelling to various parts of the empire, Marco Polo started back for Venice. Departing from Zaitun (now called Quanzhou in China), with a brief halt at Sumatra (to avoid traveling during the monsoon season), they crossed the ports of Ceylon (now Sri Lanka), and reached the Persian port of Hormuz again. From here, travelling by land through Trebizond (now Trabzon, Turkey) and Constantinople, he reached Venice in 1295.

India and China were known) and thought he could do it by sailing *westwards* from Europe.

Such a western sea route became desirable for other reasons as well. By the end of the fifteenth century, land routes to Asia from Europe were becoming increasingly dangerous due to the unavoidable encounters with hostile armies. Portuguese explorers solved this problem by taking to the sea, sailing south along the West African coast and around the Cape of Good Hope, circumnavigating Africa. (Incidentally, this route is ancient. The first person to circumnavigate Africa seems to have been the Phoenician navigator, Hanno, around 500 BC! Hanno noticed that, at the southern end of Africa, the noonday Sun was in the north. This was ridiculed by Herodotus, the Greek historian, who reported the story, but of course, Hanno would never have been able to imagine the Sun shining in the 'wrong' part of the sky if he had not seen it. Columbus thought that Asia could be reached by sailing west across the Atlantic.

Box 9.1: Marco Polo on Chinese Paper Money

When ready for use, it [a specially prepared paper] is cut into pieces of money of different sizes [. . .] The coinage of this paper money is authenticated with as much form and ceremony as if it were actually of pure gold or silver, for to each note a number of officers, specially appointed, not only subscribe their names, but affix their signets also; and when this has been done [. . .] the principal officer, deputed by his majesty, having dipped into vermilion the royal seal committed to his custody, stamps with it the piece of paper, so that the form of the seal remains impressed upon it, by which it receives full authenticity as current money, and the act of counterfeiting it is punished with death. When thus coined in large quantities, this paper currency is circulated in every part of the grand Khan's domains, nor dares any person, at the peril of his life, refuse to accept it in payment. All his subjects receive it without hesitation, because, wherever their business may call them, they can dispose of it again in the purchase of merchandise they may have occasion for, such as pearls, jewels, gold or silver.

From: The Travels of Marco Polo, Everyman's Library, New York and London, 1950.

It is a popular myth that Columbus knew the Earth was round, while everyone else thought it was flat. This is far from the truth — the European scholars of the time had accepted that the Earth was round and seasoned sailors certainly knew this. What prompted Columbus to undertake his voyage, however, was an interesting calculational error.

As early as the second century AD, Ptolemy ($\sim$ 100–170) had drawn up a map of the known world and given estimates of the distances between various points on the globe. This map (Fig. 9.3) and later versions made from it claimed the length of one degree to be about 56.6 Italian miles, with a mile being about 1477 metres. Such a conversion made the equator smaller by about one quarter. A map, produced with this error, by an Italian map-maker called Toscanelli came into Columbus's possession. Columbus calculated the distance between Spain and India over land to be 282 degrees and the distance over the sea to be about 78 degrees. Using the (wrong) conversion of degrees into miles, he estimated that India should be about 3900 miles from the Canary Islands in the Atlantic, which is more or less where America happens to be!

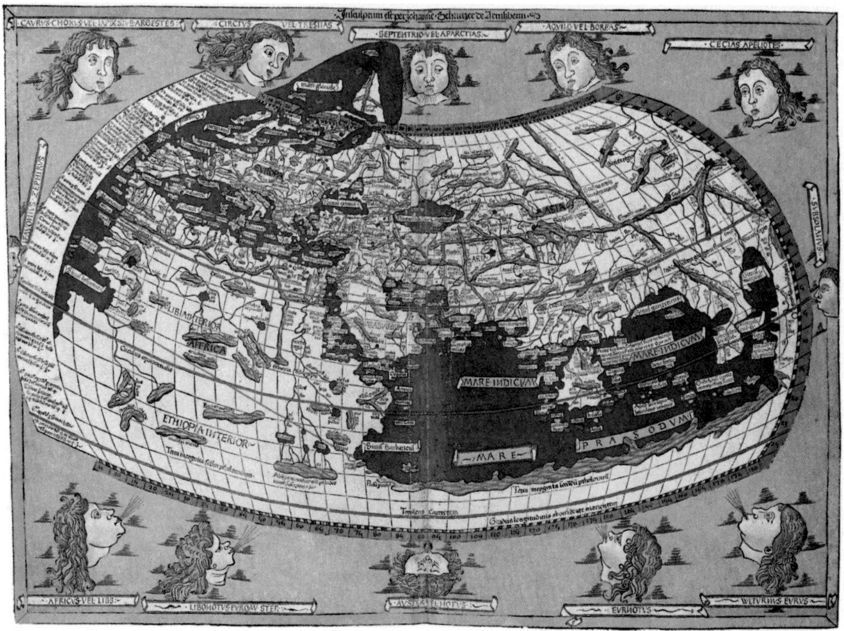

Fig. 9.3: Ptolemy's map of the world which originally appeared in his second century work *Geography* [12]. The Greek manuscript, discovered in the thirteenth century, led to further reproductions. This original map, and its later versions, claimed the length of one degree to be about 56.6 Italian miles. Such a conversion made the Earth smaller in size than it really is, and all distances became too short. Just such an erroneous map, made by an Italian map-maker Toscanelli, came into Columbus's possession, causing him to (under)estimate the distance by sea between Spain and India as about 3900 miles, which is more or less where America happens to be!

Columbus tried to get many kings and nobles to finance his trip, but met with a series of failures. In fact, it is interesting that the Portuguese king, John II referred the project to geographers, who actually pointed out that there must be something wrong with the maps Columbus was using. They felt that the definitive route to Asia was around the southern tip of Africa and that, by going west, Columbus was going the wrong way. Of course, they were quite right but what they (or Columbus or anyone else!) did not know was that between Europe and Asia lay an unknown continent (the American continent), roughly about 3600 miles away.

Columbus finally managed to get some funds from Spanish royalty — Ferdinand and Isabella — and sailed from Spain on 3 August 1492. He landed on an island which he assumed was in Asia, and named it San Salvador (present-day Bahamas) on 12 October 1492. After months of futile exploration of the neighbouring islands (present day Caribbean) in search of "pearls, precious stones, gold, silver, spices of the East", he returned back to Barcelona in March, 1493, leaving behind some of his men in a makeshift settlement on Hispaniola (present day Haiti).

Interestingly enough, on his return from the first voyage, Columbus and his remaining crew came home to a hero's welcome. He showed off what he had brought back from his voyage to the royalty, including small samples of gold, pearls, gold jewellery stolen from natives, and even a few natives he had kidnapped. He gave the monarch a few of the gold nuggets, gold jewellery, and pearls, as well as the previously unknown tobacco plant, the pineapple fruit, and the turkey. However, he did not bring any of the coveted East Indies spices, such as the exceedingly expensive black pepper, ginger, or cloves. In his log, he wrote "there is also plenty of 'aji', which is their pepper, which is more valuable than black pepper, and all the people eat nothing else, it being very wholesome" [3]. The word 'aji' is still used in South American Spanish for chili peppers.

Columbus's claims were indeed extravagant. He insisted he had reached Asia and an island (Hispaniola) off the coast of China. His descriptions were part fact, part fiction: "Hispaniola is a miracle. Mountains and hills, plains and pastures, are both fertile and beautiful [. . .] the harbors are unbelievably good and there are many wide rivers of which the majority contain gold. [. . .] There are many spices, and great mines of gold and other metals [. . .]" But this happy circumstance of the hero status for Columbus did not last.

He started his second voyage in September 1493. Finding that his Hispaniola settlement had been destroyed, he left his brothers to rebuild it and sailed further west, still in search of the wealth of the Indes. He now landed on an island (named by him as Dominica) and took back natives as slaves, much to their horror, in lieu of the material riches he had promised to his monarch in Spain. On 30 May

1498 Columbus embarked on his third voyage and landed on Trinidad and the South American mainland. When he returned to the Hispaniola settlement, he found that the colonists had staged a bloody revolt against the Columbus brothers' mismanagement and brutality. The situation became so bad that the Spanish authorities had to send a new governor to take over. Christopher Columbus was in fact arrested and returned to Spain in chains.

In 1502, the ageing Columbus managed to get himself cleared of most of the serious charges and pleaded with the Spanish king to pay for one more trip across the Atlantic. Under strict orders from the royalty not to stop at Hispaniola, Columbus travelled all the way to Panama — just miles from the Pacific Ocean — where he had to abandon two of his four ships due to an attack by hostile natives. Empty-handed and disappointed, the aged explorer returned to Spain in 1504 and died in 1506.

Columbus's voyage caught people's imagination in Europe, and the age of exploration got into full swing. It is certainly true that there were great sailors before Columbus — for instance, the Portuguese prince, Henry the Navigator (1394–1460), even set up a school of navigation in Portugal — but Columbus's voyage injected some much-needed glamour into exploration of the sea. The fact that the new lands he discovered contained vast mineral wealth and riches — which could be plundered rather easily — was another motive for the explorations; indeed it was this greed which really prompted the kings and nobles to finance the expeditions in the first place.

One such interesting after-effect of the voyages of Columbus was the following. In 1496, when Columbus returned from one of his trips, he was met by Amerigo Vespucci (1454–1512), a businessman from Florence, at Seville [4]. This meeting greatly encouraged Vespucci to undertake several voyages similar to those of Columbus, again funded by Ferdinand and Isabella of Spain. Vespucci was essentially enticed by the prospect of fame and glory and actually left his business behind to become an explorer.

On 10 May 1497, he started on his first voyage, and most probably there were three more. Two of these (the second and third) seem certain and are confirmed from other sources. The evidence for the other two comes from letters containing an account of four voyages, but these are doubted by some historians [5]. On his third voyage, which was probably the most successful, he discovered what we now call Rio de Janeiro and Rio de la Plata.

Vespucci strongly believed — quite correctly — that he had discovered a new continent. He called it 'Mundus Novus' (New World) — what we now know as South America. During his voyages, he also discovered the mouth of the Amazon river and Trinidad. Viewing the skies of the southern hemisphere, he catalogued the stars, including Alpha and Beta Centauri and the Southern Cross constellation, which were unknown to contemporary Europeans [6].

Box 9.2: Why Do We Call it America?

Obviously, the continents are called the Americas in honour of Amerigo Vespucci. But how this came about is a rather amusing story in itself.

Vespucci's reputation during his lifetime had gone through ups and downs, and he was viewed as someone who had attempted to steal Columbus's glory. In reality, this does not seem to be the case; Vespucci does not seem to have pushed for the continent to be named after him. In fact, the name was due to the work of the German clergyman and amateur cartographer Martin Waldseemuller (1470–1520).

In 1507, Waldseemuller and some other scholars were working on producing large maps and Waldseemuller suggested that a portion of Brazil that Vespucci had explored be named 'America' (Fig. 9.4), a feminine version of Vespucci's first name. Waldseemuller's maps sold thousands of copies all over Europe. Later, in 1538, another famous map-maker Gerardus Mercator (1512–1594) used the name 'America' to refer to both the northern and southern hemispheres of the New World, and interestingly enough, this name stuck!

There have been some suggestions that, in naming the continent after Vespucci, some injustice has been done to Columbus. Actually, this does not seem to be the case. Until his death, Columbus believed that he had reached Asia and not a *new* continent. But to Vespucci, this claim looked impossible. It was clear to him that the region explored by Columbus looked nothing like the description of Asia in, for example, Marco Polo's writings. Even the pattern of stars in the sky did not match. So it was Vespucci who correctly claimed that the new land was not Asia, but a new continent totally unknown to the ancients, and that between the new continent and Asia there must exist a second ocean. This claim also had the effect of overthrowing the ancient Greek notions of the world. Because of Vespucci's recognition of this startling fact, he was honoured with the naming of the new continent as America [7].

As it turns out, that is not quite the end of the story. By 1513, when Waldseemuller and the Saint-Die scholars published the new edition of Ptolemy's 'Geographiae', and again in 1516 when Waldseemuller's famous 'Carta Marina' was printed, the name 'America' did not appear in his maps! The area named 'America' on the 1507 map was referred to as Terra Incognita (Unknown Land). South America was called Terra Nova (New World) and North America, named Cuba, was shown to be part of Asia. Maybe Waldseemuller was having second thoughts about honouring Vespucci exclusively for his discovery. However, order was restored in the later cartographic contributions by Johannes Schoner in 1515 and by Peter Apian in 1520, who adopted the name 'America' for the Western Hemisphere. The cartographer Gerardus Mercator, in 1538, used the name 'America' for both the northern and southern parts, and that name continues to this day.

The country we now know as Colombia is indeed named after Christopher Columbus, even though — rather ironically — he never set foot on Colombian soil. It was Alonso de Ojeda, one of Columbus's companions on his second voyage, who was the first European to set foot in that land, in 1499.

As could be expected, shortly after Columbus's voyage, Spain and Portugal were at loggerheads on how to share the loot from the new lands. As usual, the then Pope (Pope Alexander VI) intervened and negotiated a compromise in 1494. He drew a line in the Atlantic a hundred leagues west of the Cape Verde Islands, then 'gifted' all lands to the west of this line to Spain and all those to the east to Portugal. This arrangement was so vague that it started off a race between the two nations to control trade routes, whence the line of demarcation kept shifting in favour of one side or the other. A settlement was made between the two nations at the Treaty of Tordesillas in 1506, with a decree (again!) from Pope Julius II (who had succeeded Pope Alexander VI) and the demarcation line moving 270 leagues west of the Cape Verde Islands.

This boundary line was now in the middle of the Atlantic, roughly halfway between the Cape Verde Islands and the Caribbean Island of Hispaniola, which gave the Portuguese access to the African continent on the east and the newly discovered (jutting-out part) of Brazil in the west. However, the decree could not fix the line's location across the globe because there was no way of determining the longitude. (That had to wait for another two hundred years!) The way around the globe was unknown and so were the locations of land and oceans to the west and south of Brazil. These inadequacies were cleverly exploited by Ferdinand Magellan under

Fig. 9.4: Waldseemuller's map of the world introducing the name America [13]. For a close-up of the region (the lower leftmost panel) bearing the name, see Fig. 9.5. It was Waldseemuller who introduced this name in this map, which he made in 1507. He suggested that the part of Brazil which Amerigo Vespucci had explored should be named 'America', a feminine version of Vespucci's first name. In 1538, another famous mapmaker, Gerardus Mercator, decided to use the name 'America' for both the northern and southern parts of the New World. This is how America initially got its name.

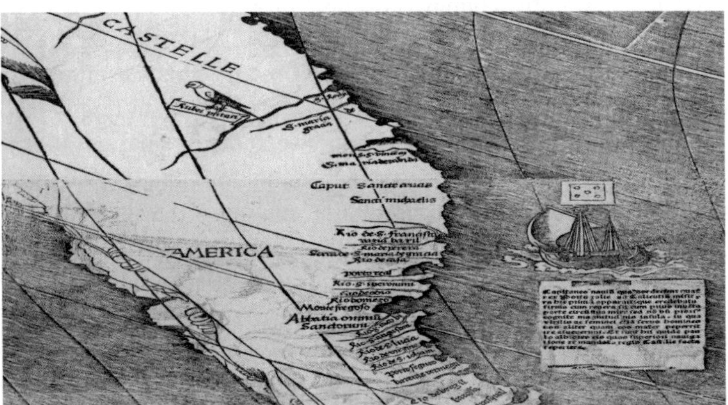

Fig. 9.5: Waldseemuller's map with the close-up of the region containing the name America [14].

Fig. 9.6: The Strait of Magellan [15]. The Strait of Magellan is a navigable sea route in southern Chile, separating mainland South America to the north and Tierra del Fuego to the south. The strait is the key natural passage between the Atlantic and Pacific oceans. Though considered a difficult navigational route due to the narrowness of the passage and unpredictable winds and currents, it is shorter and more sheltered than the often stormy route known as the Drake Passage. Along with the narrow and treacherous Beagle Channel, these were the only three sea routes between these two oceans until the construction of the Panama Canal.

the patronage of the Spanish king (though originally he was a Portuguese explorer) [8]. He intended to sail south of Brazil, along the eastern coast of what we now call South America, until the land ended and then keep sailing westwards (thereby always remaining on the correct side of the papal line) and reach the eastern end of the Spice Islands [9].

Magellan (1480–1521) was a page at the court of John II, the Portuguese king who had refused to finance Columbus. Magellan had also participated in the Portuguese expeditions to the East Indies and had joined the Portuguese army in battle in Morocco, where he was permanently lamed in action. He was later caught trading with the Moroccans — which was equivalent to treason — and was dismissed in 1517 without a pension. Bitter at this treatment, he switched loyalties, joined Spanish service, and rose high. He pointed out to Charles V of Spain that if the Spaniards kept sailing westwards they would always be on the right side of

Fig. 9.7: Magellan's ship, Victoria [16]. Details of the ship are from a map by Ortelius dated to 1590. Victoria was part of the expedition commanded by Ferdinand Magellan. Though the expedition began on 10 August 1519 with five ships, Victoria was the only ship to complete the voyage, returning on 6 September 1522. During the final leg, on 21 December 1521, Victoria sailed on from Tidore in Indonesia alone, because the other ships left the convoy due to lack of rations. The ship was in terrible shape with her torn sails, and only kept afloat by the continuous pumping of water. Victoria was later repaired, bought by a merchant shipper and sailed for almost another fifty years before being lost with all hands on a trip from the Antilles to Seville in about 1570. A life-size replica of this ship exists today in a museum in Chile's (and the world's) southernmost city of Punta Arenas.

the papal line but could nevertheless reach the Indies, which, by the papal decree, was given to the Portuguese to explore. Magellan was essentially using Columbus's idea but was doing it right, having realized that the American sub-continent was not the Indies they were seeking.

The Spanish monarch liked this idea and sponsored Magellan to set sail with five ships on 10 August 1519. The ships crossed the Atlantic, found a small passage into the Pacific Ocean (now called the 'Strait of Magellan') at the southern end of South America (Fig. 9.6), and just about managed to reach the island of Guamnear (the present Philippines) on 6 March 1521, virtually on the brink of starvation. Magellan was later killed in a quarrel with the Filipino natives, but one of his five ships, the Victoria (Fig. 9.7), managed to make its way across the Indian Ocean, around the southern tip of Africa and back to Spain, arriving there on 8 September 1522. (So, to be technically accurate, it was Juan Sebastian del Cano — a Spanish navigator who assumed command of the expedition when Magellan was killed — who became the first circumnavigator of the globe.) The Victoria returned to Spain with a shipload of spices, the value of which was greater than the cost of the entire original fleet!

This circumnavigation of the globe was crucial to prove beyond doubt three facts. First, the estimate of the size of the Earth by Ptolemy (see Fig.9.3) was wrong and — rather interestingly — the earlier estimates by Eratosthenes ($\sim$ 276–194) were right. Second, there is a single ocean — not 'seven seas' as thought by the Greeks — which girdles the Earth. And third, it showed that vast lands existed on the Earth, with new animals and plants, about which Aristotle and other 'deep' thinkers knew absolutely nothing. All this emphasized the shortfallings of accumulated ancient knowledge and prepared a fertile ground for the Copernican Revolution.

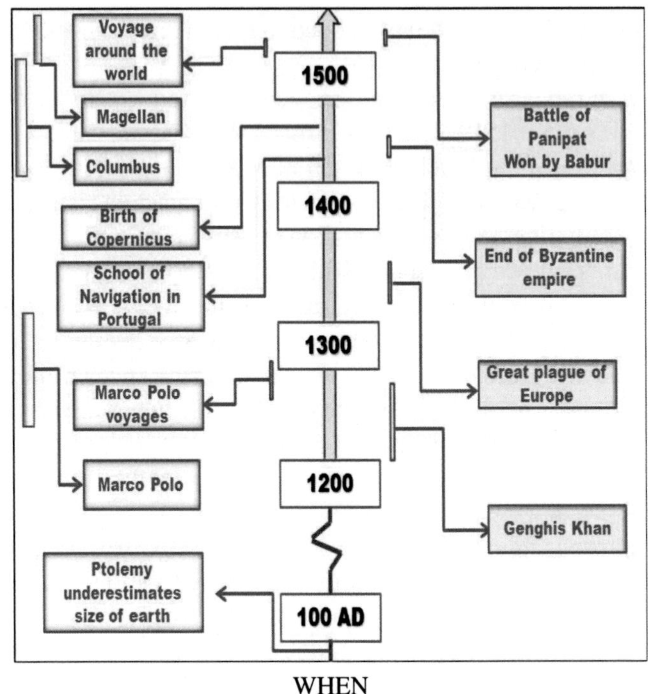

WHEN

Notes, References, and Credits

Notes and References

1. For more on Marco Polo's travels, see, e.g.,
   Otfinoski, Steven (2003), *Marco Polo: to China and back*, Benchmark Books, New York [ISBN 0-7614-1480-0].
   Bergreen, Laurence (2007), *Marco Polo: From Venice to Xanadu*, Knopf Doubleday Publishing Group, New York [ISBN 9780307267696].
   Burgan, Michael (2002), *Marco Polo and the silk road to China*, Compass Point Books, Mankato [ISBN 978-0-7565-0180-8].
2. For more on Columbus and his voyages, see:
   Phillips Jr. William D. (2012), "Columbus, Christopher", in *The Oxford Companion to World Exploration*, David Buisseret (ed.), Oxford University Press, Oxford.
   Keen, Benjamin (1978), (Translation of) *The Life of the Admiral Christopher Columbus* (by His Son Ferdinand) Greenwood Press, Westport [ISBN 978-0-313-20175-2].
3. See,

Zinn, Howard (2009), *A People's History of the United States: 1492 to Present*, p. 3, Harper Collins, New York, USA [ISBN 9780061989834].

4. For more details, see, e.g.,
Hoogenboom L. (2006), *Amerigo Vespucci: A Primary Source Biography*, The Rosen Publishing Group, USA [ISBN 978-1-4042-3037-8].
Donaldson-Forbes J. (2002), *Amerigo Vespucci*, Powerkids Pr. Publishers, USA [ISBN 978-0-8239-5833-7].
5. There is some controversy about the one called 'Lettera al Soderini' or just 'Lettera', a letter in Italian addressed to Piero Soderini. Printed in 1504 or 1505, this letter claimed to give an account of four voyages to the Americas made by Vespucci between 1497 and 1504. In the eighteenth century, three previously unpublished letters from Vespucci to Lorenzo de' Medici were rediscovered. One of these describes a voyage made in 1499–1500, which corresponds to the second of the 'four voyages'. Another was written from Cape Verde in 1501, in the early part of the third of the four voyages, before crossing the Atlantic. The third letter was sent from Lisbon after the completion of that voyage. See also:
Formisano, L. (Ed.) (1992), *Letters from a New World: Amerigo Vespucci's Discovery of America*, Marsilio Publishers, New York [ISBN 0-941419-62-2].
6. Curiously enough, these stars were known to ancient Greeks but the precession of the equinoxes had lowered them below the horizon so that they had been forgotten by this time.
7. Hebert, John R. (2003), *The Map that Named America*, Library of Congress Information Bulletin, Vol. **62**, No. 9.
8. For more on Magellan's voyages, see, e.g.,
Stefoff, Rebecca (1990), *Ferdinand Magellan and the Discovery of the World Ocean*, Chelsea House Publishers, New York [ISBN 0-7910-1291-3].
Beaglehole, J.C. (1966), *The Exploration of the Pacific*, Adam and Charles Black, London [OCLC 253002380].
9. Bergreen, Laurence (2004), *Over the Edge of the World: Magellan's Terrifying Circumnavigation of the Globe*, Harper Perennial, New York [ISBN 978-0066211732].

Figure Credits

10. Figure 9.1 courtesy: Wikimedia Commons,
https://commons.wikimedia.org/wiki/File:Marco_Polo_at_the_Kublai _Khan.JPG (from public domain).
11. Figure 9.2 courtesy: Canuckguy BlankMap-World6.svg, Wikimedia Commons, https://commons.wikimedia.org/wiki/File:Polo%27s_journey.svg (from public domain).

12. Figure 9.3 courtesy: Wikimedia Commons, https://commons.wikimedia.org/wiki/File:Claudius_Ptolemy-_The_World.jpg (from public domain).
13. Figure 9.4 courtesy: Martin Waldseemuller, Wikimedia commons, https://commons.wikimedia.org/wiki/ File:UniversalisCosmographia.jpg. (from public domain).
14. Figure 9.5 courtesy: Martin Waldseemuller, Wikimedia commons, https://commons.wikimedia.org/wiki/ File:Waldseemuller_map_closeup_with_ America.jpg (from public domain).
15. Figure 9.6 courtesy: Wikimedia Commons, https://en.wikipedia.org/wiki/File:Strait_of_Magellan.jpeg (from public domain).
16. Figure 9.7 courtesy: Wikimedia Commons, https://en.wikipedia.org/wiki/File:Detail_from_a_map_of_ Ortelius_-_Magellan%27s_ship_Victoria.png (from public domain).

10

The First Steps of Modern Medicine

The Dark Ages ($\sim$ 500–1200 AD) in Europe was marked by man's preoccupation with theology and the afterlife. Though Christian monasteries did act as a mute reservoir of knowledge, translating the ancient Greek works on medicine, the prevalent view was that a disease was a punishment for sin and should be cured by prayer and repentance. It was also maintained that the human body was sacred and the dissection of a corpse was a sin; any point of view contrary to the establishment views was treated as blasphemous and referred to the Inquisition. All this did great harm to the advancement of medical science.

Needless to say, medical science could not progress without an accurate knowledge of the anatomical structure of the human body. A major barrier to such knowledge came from the social and religious attitudes towards dissection of the human body. "In the recesses even of the rational mind," says British historian Ruth Richardson [1], "there lurked the fear that the mutilation of the corpse might have eschatological implications."

And indeed, different civilizations took very different attitudes towards this issue. In the fourth century BC — the era of Aristotle (384–322 BC) and Hippocrates (460–370 BC) — the prevailing religious attitudes considered the body and soul to be separate entities, thereby making dissection acceptable. Galen ($\sim$ 129–200 AD), for example, is supposed to have said that Alexandria, where he had studied, was the only place where anatomy could be learned, possibly for this reason. But during the time of the Romans, this practice was banned and — for the next 1000 years or so — scientists were reduced to dissecting only animals.

T. Padmanabhan and V. Padmanabhan, *The Dawn of Science*,
https://doi.org/10.1007/978-3-030-17509-2_10

So it was only by the end of the thirteenth century that there was a slow resurgence of interest in the works of the ancient Greeks — both scientific and non-scientific. Scholars tried to follow the ancient lines of thinking and out of this arose modern Western science. An interest in humanities led — somewhat indirectly — to the study of human anatomy again, and two major Italian schools of medicine came into being, at Salerno and Bologna.

Human dissection started again in the thirteenth century in the University of Bologna, one of the great medieval universities. Several incidents in the early thirteenth century set the stage for diluting the taboo against dissections. First, Emperor Frederic II had issued an imperial decree in 1238 AD, authorizing the performance of dissections on the bodies of executed criminals for academic purposes. Second, there was the practice of dismembering the bodies of the Crusaders so that the bones could be carried back to their families in Europe for burial — which made anatomical dissections less of an issue. Third, historians claim that there was also an interest amongst the legal scholars of the University of Bologna, which had a famous law school, for the resumption of human dissection in order to ascertain causes of death [2]. All these helped the cause.

One of the earliest anatomists was Mondino de Luzzi (1270–1326), who hailed from Bologna [3]. In his time, the dissection of corpses — even when attempted — was considered to be an act beneath the dignity of a teacher. So the medical man would lecture from a high podium while some lowly servant would dissect a corpse to illustrate the points (see Fig. 10.1). It was, of course, impossible to coordinate the activities of the two; what was more, the servant — not being a man of science — could only make sloppy dissections.

Mondino changed this practice and started dissecting corpses himself. His lectures were consequently much more popular and soon earned him the title 'Restorer of Anatomy'. In 1316, he wrote his major work, *Anothomia corporis humani*, the first textbook which dealt exclusively with anatomy. It was based on his own dissection of humans. This book was published at Padua in 1487 and went through 39 separate editions and translations.

Mondino's work was taken forward by his students, and dissection became more commonplace. A public dissection occurred at Padua in 1341; later, they were decreed at the University of Montpelier in 1366, at Venice in 1368, and at Florence in 1388. In Padua, a theatre for dissections was erected in 1445.

Unfortunately, the tradition started by Mondino did not last. His successors went back to the older practices and continued to hold the earlier writings of Aristotle and Avicenna (980–1037) to be sacred, hence embodying the ultimate truth. It took nearly two more centuries for this stranglehold to loosen.

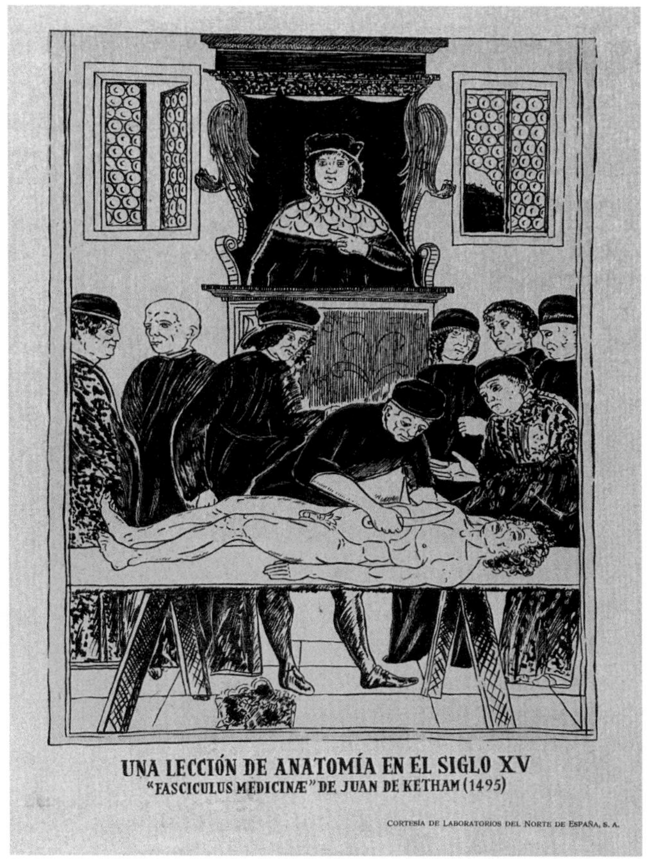

Fig. 10.1: Before the days of Vesalius (1514–1564), the anatomy teacher never did the dissections himself; while the teacher lectured, lowly servants, who had little or no medical knowledge, dissected a human corpse. This drawing is based on a woodcut made in 1493 [7].

The major contributors to this 'revival of reason' in medical science were the Swiss physician and alchemist Paracelsus (1493–1541) and the Flemish anatomist Andreas Vesalius (1514–1564) — two men who were very different in temperament, talent, and outlook.

Paracelsus's original name was Theophrastus Bombastus von Hohenheim and, obviously, you can't blame him for wanting to change it! But the choice of name

Fig. 10.2: Portrait painting [8] of Paracelsus (1493–1541). He changed his original name — Theophrastus Bombastus von Hohenheim — to Paracelsus, which means 'better than Celsus'. He was an outspoken and charismatic lecturer of medicine at the University of Basel. Among other contributions, he also wrote the first book (on 'miners' sickness') dealing with what we now call occupational diseases.

portrays his character very well: Paracelsus means 'better than Celsus'. Celsus was the famous Roman physician whose works had just been translated around that time and had made a tremendous impact on contemporary thinkers. Paracelsus, in fact, thought that he was better than everybody and did not hesitate to say so [4]. At a very early age, he acquired a good knowledge of mining, metallurgy, and chemistry, and he was interested in pursuing the alchemist's goal of turning lead into gold. He also acquired medical education at several universities in Europe, moving from place to place, often because of the fights he invariably got involved in wherever he happened to be. He soon realized that chemistry could play a useful role in producing medicines to cure diseases. In a way, this also marked the beginning of the transition from alchemy to chemistry.

Paracelsus returned home in 1524, after nearly 10 years of wandering, and was appointed the town physician and lecturer on medicine at the University of Basel. He began his lecture series by publicly burning the works of Galen and Avicenna at the entrance to the university. Though the authorities were quite infuriated, the students cheered the eccentric and charismatic teacher. His teachings were a curious mixture of deep insights and stupid blunders. For example, he developed a clear clinical description of syphilis and treated cases with carefully measured doses of mercury compounds. He understood that the 'miner's disease' (silicosis) was caused by the inhalation of metal vapours and not by the anger of the mountain spirits. He also wrote the first book ('Miners' Sickness') dealing with what we now call occupational diseases; cleanliness, he emphasized, was crucial for health. He produced scores of useful medical compounds containing mercury, sulphur, iron, and copper sulphate. Yet, while doing all this, he firmly believed in the four elements of the Greeks and went on assiduously searching for the elixir of life!

Box 10.1: Habeas Corpus?

The problems for progress in anatomy were not confined to the social and religious taboos on human dissections. Even after overcoming these taboos, for centuries physicians had to deal with the dearth of human corpses!

In almost all cultures, the afterlife of the soul was thought to be affected by what happens to the body after death and hence people did not look kindly upon their bodies being violated after death. (Much worse, while just presumed dead. Folklore has it that Vesalius was once very surprised, while starting a dissection in Spain, by a cry of "Ouch!" from the supposedly dead body!)

Looking for dead bodies, scientists in Europe found themselves coming in contact with all sorts of shady characters, including grave robbers. Henry VIII enacted a law to donate the bodies of executed criminals to medical schools; but even then the supply could not meet the demand, leading — as usual — to the development of a profitable black market. There was a case of William Burke and William Hare in Scotland (in the eighteenth century), who ran a boarding house which was really a cadaver factory. They murdered their tenants in cold blood and delivered the bodies to anatomy schools for about 10 pounds each — which, in those days, was nearly five months wages for the average working class man.

By the mid-1830s, many US states had passed the Anatomy Act, which ended the practice of providing the corpses of criminals and replacing them with those of people who were too poor to afford a regular burial.

Unfortunately, this did not stop the grave-robbing, and, of course, the poor resented the act. By and large, doctors and governments found that there was no simple solution to the problem.

Slowly, and only over a long period of time, the public accepted the need for anatomical dissections — and hence the need for corpses — for the progress of medicine. Attitudes towards death and its aftermath changed over a span of time and medical schools began to receive a steady supply of legitimately donated corpses.

Vesalius's life, on the other hand, was rather prosaic in comparison but more fruitful [5]. Born in a wealthy and powerful family, he studied at a medical school in Paris from 1533–1536. As part of his education, he made a detailed study of Galen and Rhazes (854–925) and, at first, could find no fault with their writings. On receiving his degree from the University of Padua, he was appointed as a lecturer in surgery. While preparing for his lectures, he spent a fair amount of time dissecting corpses (see Fig. 10.3). It then became clear to him that Galen and others had not based their writings on the actual dissection of the human body (which was indeed banned in Roman times), but on extrapolations from studies of animal bodies. He had the courage to declare his own conclusions openly and proceeded to write a textbook on human anatomy that was based on correct knowledge.

This book, *De Humani Corporis Fabrica Libri Septem* (Seven Books on the Structure of the Human Body), was superbly illustrated — most probably at the studio of the renaissance artist, Titian — and was printed in 1543 (see Fig. 10.4). It represented the culmination of several important trends in medicine, especially the revival of ancient learning and the growth of attention to the human body as an object of serious study. *De Fabrica* also contained an extensive and accurate description of the human body and provided anatomy with a new language for its description. Indeed, one might say that anatomy came of age with Vesalius. (Curiously enough, *De Fabrica* was published in the same year as Copernicus (1473–1543) published his book outlining the heliocentric theory — another milestone in science, which we will describe in Chap. 11.)

Vesalius was quite accurate as an anatomist, but his ideas on other branches of medicine were not always so precise. He did believe (quite correctly) that the brain and the nervous system represented the seat of emotions — thereby categorically denying the earlier Aristotelian view that these functions were governed by the heart. This idea, fortunately, gained acceptance. At the same time, he believed in

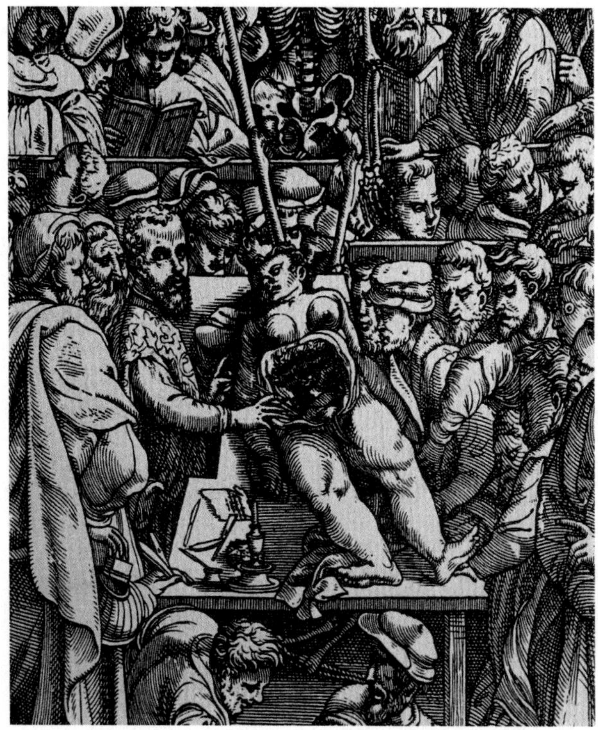

Fig. 10.3: Anatomical dissection by Andreas Vesalius of a female cadaver, attended by a large crowd of onlookers, based on a woodcut [9] from 1555. Vesalius did the dissections himself, which led to significant advances in anatomical knowledge. His book, *De Humani Corporis Fabrica Libri Septem*, contained an extensive and accurate description of the human body, and provided anatomy with a new language for its description. In fact, anatomy came of age with Vesalius.

Galen's view on the circulation of blood and thought that blood passes from one ventricle to another, within the heart, through some mysterious process.

Following the publication of his book in 1543, Vesalius decided to relinquish his anatomical studies for the practice of medicine. With the long tradition of imperial service in his family, he took a position as Imperial Physician to Emperor Charles V. This was not a wise decision because much of his time was devoted to the complaints of the gluttonous Emperor and, as Vesalius wrote,"to the Gallic disease, gastrointestinal disorders, and chronic ailments, which are the usual complaints of my patients" [6]. He travelled with the court, treating injuries from battle, performing post-mortems, administering medications, etc. Imperial service, once entered, could not be abandoned; so Vesalius remained the Emperor's physician for thirteen years, until the latter's abdication, after which he entered the service of Charles's son, Philip II, and moved to the new imperial court in Spain. This move was even more of a mistake than taking up royal service in Charles's court in 1543! It was difficult even to get hold of a skull in Spain and he longed to go back to Padua [5, 6].

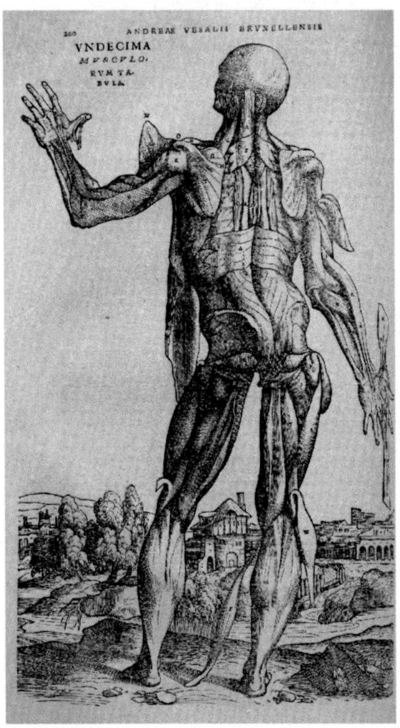

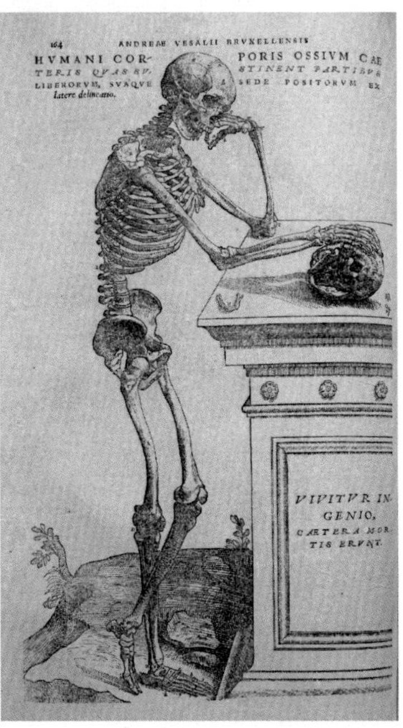

Fig. 10.4: Two images from Andreas Vesalius's *De Humani Corporis Fabrica* (1543). Left: A page showing the muscle groups of the human body [10]. Right: A woodcut from the book shows a skeleton meditating over a skull [11]! The inscription on the stone plinth reads, "Vivitur ingenio, caetera mortis erunt." (Genius lives on, all else is mortal.) This book was very well illustrated and became a medical classic. It has been suggested that the illustrations were done at the studio of the renaissance artist, Titian.

Unfortunately, when he finally did get a chance to resume the professorship at Padua, fate intevened. On a return trip from a pilgrimage to Jerusalem, intending to accept the professorship at Padua, he was shipwrecked on the island of Zakynthos off the Greek coast. He died there, apparently in debt, and was buried somewhere in the island.

Vesalius met this unfortunate end at the relatively young age of 50. Had Vesalius resumed his researches at Padua, it is quite possible that medical science would have surged ahead by many years, and medical history would have been very different.

Box 10.2: The Last Pilgrimage of Vesalius

In 1564, Vesalius embarked on a pilgrimage to the Holy Land, sailing with the Venetian fleet, via Cyprus. When he reached Jerusalem he received a message from the Venetian senate requesting him to accept the Paduan professorship, which had now become vacant on the death of his friend and pupil Fallopius (1523–1562). His return voyage, however, was far from smooth. After struggling for many days in the Ionian Sea, he was shipwrecked on the island of Zakynthos. He died there, in such debt that a benefactor had to step in and pay for his funeral. He was only fifty years of age and was buried somewhere on the island of Zakynthos.

For many years it was rumoured that Vesalius's pilgrimage was due to the pressures from the Inquisition. Modern biographers today dismiss this rumour as being without foundation. It appears that this story was spread by Hubert Languet, a diplomat under Emperor Charles V, who claimed in 1565 that Vesalius had performed an autopsy on an aristocrat in Spain while the heart was still beating. This is supposed to have led the Inquisition to condemn him to death. It is claimed that Philip II had the sentence commuted to a pilgrimage. The story resurfaced several times over the next few years, living on until recent times, but seems to be without foundation. A more likely scenario seems to be that the journey was related to an illness, real or feigned, and used as an opportunity to leave Spain [6].

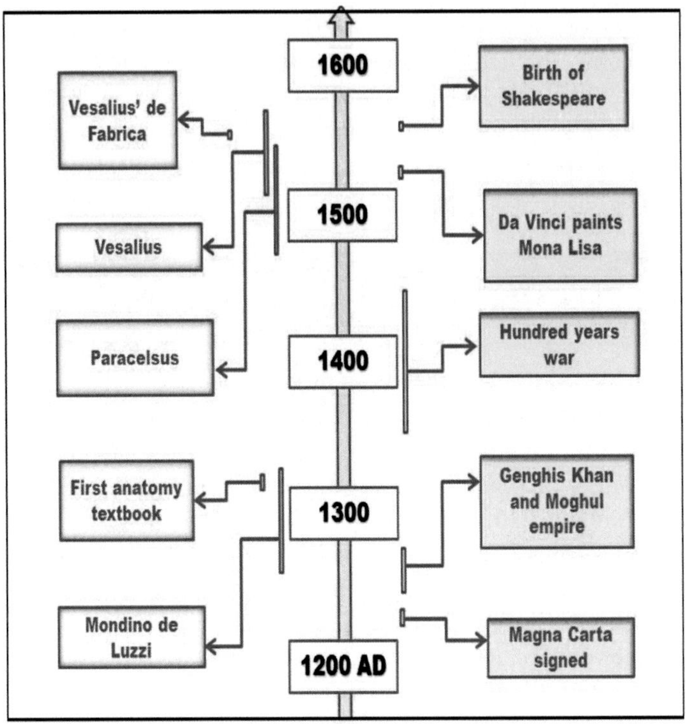

WHEN

Notes, References, and Credits

Notes and References

1. Richardson, Ruth (2001), *Death, Dissection, and the Destitute*, University of Chicago Press, Chicago [ISBN 978-0226712406].
2. Walker H.K., Hall W.D. and Hurst J.W., (Eds) (1990), *Clinical Methods: The History, Physical, and Laboratory Examinations*, Butterworths, Boston [ISBN 978-0409900774].
3. For the history of medicine during this period, see:
   Olmi, Giuseppe (2006), *Representing the body – Art and anatomy from Leonardo to Enlightenment*, Bologna University Press, Bologna.
   Singer, Charles (1957), *A Short History of Anatomy from the Greeks to Harvey*, Dover, New York.

Siraisi, Nancy (1990), *Medieval and early Renaissance medicine: an introduction to knowledge and practice*, University of Chicago Press, Chicago [ISBN 978-0226761305].

4. For more on Paracelsus:
   Pagel, Walter (1982), *Paracelsus: An Introduction to Philosophical Medicine in the Era of the Renaissance*, Karger Publishers, Switzerland [ISBN 3-8055-3518-X].
   Webster, Charles (2008), *Paracelsus: Medicine, Magic, and Mission at the End of Time*, Yale University Press, Yale [ISBN 978-0300139112].
   Ball, Philip (2006), *The Devil's Doctor*, Arrow Books, Random House, UK [ISBN 978-0-09-945787-9].
   Stoddart, Anna (2011), *The Life of Paracelsus*, Balefire Publishing, US [ISBN 978-1402142765].
   Jacobi, Jolande (1995) *Paracelsus: Selected Writings* (ed. Norbert Guterman), Princeton University Press, Princeton [ISBN 978-0691018768].
5. For more on Vesalius, see:
   Dear, Peter (2001) *Revolutionizing the Sciences: European Knowledge and Its Ambitions*, 1500-1700, Princeton University Press, Princeton [ISBN 978-0691088594].
   Debus, Allen, (1968) (ed) *Vesalius — Who's Who in the World of Science: From Antiquity to Present*, Western Co., Hanibal [ISBN 978-0837910017].
   O'Malley, C. D (1965), *Andreas Vesalius of Brussels, 1514–1564*, University of California Press, Berkeley [ISBN 978-0520009547].
6. Saunders, J. B. deCM. and O'Malley, C.D. (1973), *The Illustrations from the Works of Andreas Vesalius of Brussels*, Dover Publications, USA [ISBN 978-0486209685].
   O'Malley, C.D. (1964), *Andreas Vesalius 1514–1564: In Memoriam*, Medical History **8** (4), 299–308.

Figure Credits

7. Figure 10.1 courtesy: Wellcome Library London no.24083i .
   https://commons.wikimedia.org/wiki/File:A_dissection_in_progress;_the_anatomy_professor_at_his_lecte_Wellcome_V0010408.jpg (published under CC by 4.0).
8. Figure 10.2 courtesy: Quentin Matsys / Wikimedia Commons /Public Domain.
   https://commons.wikimedia.org/wiki/File:Paracelsus.jpg (from public domain).
9. Figure 10.3 courtesy: https://commons.wikimedia.org/wiki/File:Anatomical_dissection_by _Andreas_Vesalius_of _a_female_Wellcome_V0010428.jpg Photo number: V0010428 from Wellcome Images, made available under the Creative Commons Attribution 4.0 International license.

10. Figure 10.4 (left) courtesy: Image from Andreas Vesalius's De humani corporis fabrica (1543), page 200. Wikimedia Commons /Public Domain https://commons.wikimedia.org/wiki/File:Vesalius_Fabrica_p200.jpg (from public domain).
11. Figure 10.4 (right) courtesy: Image from Andreas Vesalius's De humani corporis fabrica (1543), page 164. Wikimedia Commons /Public Domain https://commons.wikimedia.org/wiki/File:Vesalius_Fabrica_p164.jpg (from public domain).

11

Making the Earth Move

> "In the year 1500, Europe knew less about Nature than Archimedes, who died in 212 BC, did."
>
> — A.N. Whitehead, (English historian of science, [1])

On 24 May 1543, Nicholas Copernicus (1473–1543) lay in bed, dying of a brain hemorrhage. It is said that a copy of his book *De Revolutionibus Orbium Coelestium* (On the Revolution of the Heavenly Bodies) — the publication of which he himself had delayed by nearly 30 years — was brought to his deathbed so that he could have a glimpse of it. In this book, he had detailed a system of the heavens with the Sun at the center and the planets going around it in fixed orbits. Copernicus had, in effect, stopped the Sun and set the Earth in motion [2].

Behind this event lies one of the most fascinating stories in the entire history of science. A story of extraordinary blunders, irresponsible acts, and the damaging effects of religious suppression of science. To see this in proper perspective, we have to go back in time 1800 years.

Around the third century BC, there lived a Greek astronomer, Aristarchus ($\sim$ 310–230 BC), who wrote a treatise *On the Sizes and Distances of the Sun and the Moon* [3]. In this treatise, he proclaimed clearly that the Sun — and not the Earth — was at the center of our world and that all the planets revolved around the Sun. His book became a classic of antiquity and he was indeed considered one of the foremost astronomers of his time. Both Archimedes (third century BC)

T. Padmanabhan and V. Padmanabhan, *The Dawn of Science*,
https://doi.org/10.1007/978-3-030-17509-2_11

and Plutarch (first century AD) knew about his work and mentioned him. "For, Aristarchus supposed that the fixed stars and the Sun are immovable but the Earth is carried around the Sun in a circle ..." said Archimedes; "... that the heaven is at rest but the Earth revolves in an oblique orbit while it also rotates about its own axis," is the reference in Plutarch to the ideas of the Greek astronomer.

Incredibly enough, the world forgot Aristarchus! The geocentric system of Ptolemy (100–170 AD), a much more complicated and aesthetically unappealing one, held sway even as early as the second and third centuries AD. Afterwards, through the Dark Ages, there was no hope of revival. Later, when Europe went through the Renaissance, one would have hoped for the right ideas to re-emerge. But the strong religious dogmas and theological interpretations of Aristotle's outdated ideas totally suppressed the truth for centuries. Though it was logically only one step from Aristarchus to Copernicus (or, for that matter, from Hippocrates to Vesalius or even from Archimedes to Galileo), it took centuries in Europe for this step to be taken. As A.N. Whitehead, an English historian of science, remarks: "*In the year 1500, Europe knew less about Nature than Archimedes, who died in 212 BC, did.*" So strong was the effect of dogmatic ideas on the growth of science.

The way Copernicus ended up writing his book is yet another interesting tale. Born in 1473 in Torun in Eastern Poland, Copernicus lost his father early in life. Thereafter, his uncle brought him up and gave Copernicus an excellent education. In 1496, Copernicus moved to Italy and studied medicine and canon law for 10 years. This was when he got interested in astronomy!

In those days, the positions of the planets were calculated using the system developed by Ptolemy (see Fig. 11.1). Though detailed, because of its complexity (and painful mathematics), this system was cracking at the edges. The predicted positions of the planets were getting to be too far away from the observed ones, in spite of several ad-hoc corrections introduced by later astronomers. It occurred to Copernicus that the calculations could be considerably simplified if one adopted the heliocentric system — that is, one in which the Sun is at the center.

Copernicus's genius was in putting this idea into practice and meticulously working out the details of the new model. He relied on the observations of others, because he was not good in this. (His instruments were actually less accurate than those used in Alexandria 2000 years earlier; he even made a mistake in the terrestrial coordinates of his observatory!)

Copernicus immediately realized that his model could explain several things which Ptolemy's couldn't. It must certainly have occurred to him that his was probably the correct model — with a moving Earth and all that went with it (see Fig. 11.3). But he hesitated to publish it, knowing full well that he could

Schema huius præmissæ diuisionis Sphærarum.

Fig. 11.1: In the pre-Copernican universe, the Earth is at the center, and the Sun, Moon, planets, and stars go around the Sun in concentric circles [7]. The figure is based on a diagram of the celestial orbs in the 1539 work *Cosmographia* by Peter Apian. (The positional relationship between the various spheres has been maintained in the reproduction.) The scheme involved a division of spheres made of the empyrean (fiery) heaven, the dwelling of God and all the chosen; the Tenth heaven was the first cause and the Ninth heaven was supposed to be crystalline. Thereafter, the Eighth heaven was of the firmament followed by the heavens numbered 7 to 1, giving the abodes of Saturn, Jupiter, Mars, Sun, Venus, Mercury, and the Moon.

get into trouble with the Church. A private manuscript, which he circulated himself, created considerable interest among European scholars, and in the end the German mathematician, Georg Rheticus (1514–1574), who was a fan and student of Copernicus, persuaded Copernicus to publish it as a book [4]. Rheticus also suggested the rather clever idea of dedicating the book to Pope Paul III in order to pre-empt opposition from the Church. Copernicus agreed and entrusted Rheticus with this task.

Fig. 11.2: Portrait of Copernicus [8], who, in his work, *De Revolutionibus Orbium Coelestium* (On the Revolution of the Heavenly Bodies), described a system of the heavens with the Sun at the center and the planets going around it in fixed orbits. This heliocentric model was, in fact, first introduced 1800 years earlier by Aristarchus, who lived in the third century BC! He wrote a treatise *On the Sizes and Distances of the Sun and the Moon* stating clearly that the Sun — and not the Earth — was at the center and that all the planets revolved around the Sun. This treatise was a classic in ancient times, and both Archimedes and Plutarch mention it in their work. Tragically for science, *the world forgot Aristarchus!* The geocentric system of Ptolemy, a much more complicated one, held sway even as early as the second and third centuries AD (mostly due to Aristotle's influence) and resurfaced after the Renaissance. So, poor Copernicus had to do it all over again!

Unfortunately, Rheticus had to leave the town, and he entrusted the charge of printing to a Lutheran minister, Osiander (1498–1552). Earlier, Martin Luther (1483–1546) had expressed strong opposition to Copernicus — which goes to show that religious reformists may not always help the progress of science — and Osiander decided to play it safe and publish the book with the addition of a — highly damaging — preface. The preface essentially conveyed the idea that the system described in the book was only a mathematical scheme and might not represent reality. It explicitly stated that "... these hypotheses need not be true or

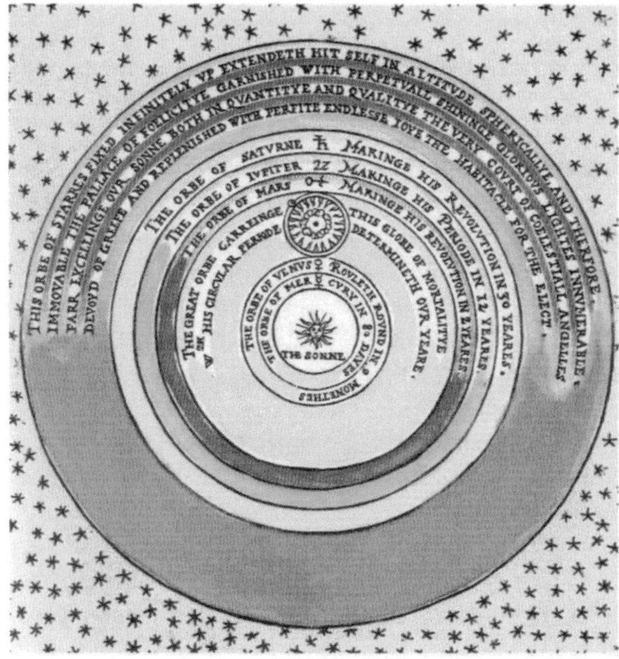

Fig. 11.3: The Copernican universe (as depicted by Thomas Digges in 1576), with the Sun at the center and the Moon going around the Earth. The stars are not located on a fixed sphere but spread out to infinite distances [9].

even probable." Historians are unsure whether Copernicus approved of this preface and the issue has not been resolved. It shouldn't be surprising if he actually did, for old Nicholas always knew which side of the bread was buttered; for one thing, nowhere in the book is there any note of thanks to Rheticus — a fact which deeply hurt Rheticus. The book finally appeared with just such a preface and a dedication to the Pope.

Box 11.1: Clues, Cover-ups, and the Climax

There were several tell-tale signs in the behaviour of heavenly bodies which clearly suggested a heliocentric world. Unfortunately, Ptolemy and his followers were not courageous enough to follow up these clues.

To begin with, the estimate of the size of the Sun (relative to the size of the Earth) had been available ever since the time of Aristarchus (around the third century BC) [3]. Though the actual figure he obtained was wrong, it was obvious that the Sun was considerably bigger than the Earth. It was thus rather strange to think that the Sun was going round the Earth instead of the other way around.

Further, the observations of the trajectories of planets revealed some strange behaviour. The planets Mercury and Venus were always seen close to the Sun, just after sunset or just before sunrise, and were never seen overhead at night. The other three planets, Mars, Jupiter, and Saturn, showed an irregular pattern of motion every once in a while. They travelled in one direction for some time, stopped in their tracks, and then moved backwards (called retrograde motion)! These features would be difficult to understand in a natural manner if the planets were moving around a stationary Earth.

It was also known that the brightness and the apparent size of the planet Venus varied periodically — something which should not happen if Venus were orbiting the Earth at a constant distance.

All these problems immediately disappear if we assume that the planets revolve around the Sun in the order Mercury, Venus, Earth, Mars, Jupiter, and Saturn. Because the orbits of Mercury and Venus are closer to the Sun than that of the Earth, they can never appear overhead at night. The retrograde motions of the other three planets are also easy to understand: if the Earth revolves around the Sun at a faster rate than these three planets, then the Earth will 'overtake' them every once in a while. When we observe from the Earth, the outer planets will appear to go backwards (see Fig. 11.4). Further, since the distance between the Earth and Venus varies quite a lot, the appearance of this planet will be altered periodically.

The Greeks had all these pieces of the puzzle, but refused to put them together. So deep-rooted and dogmatic were Aristotle's notions of circular motion that Ptolemy actually ended up saying: "We believe that the object of the astronomer [. . .] is this: to demonstrate that all the phenomena in the sky are produced by means of uniform and circular motions." To achieve this, Ptolemy had to invoke a complicated system of epicycles, in which heavenly objects moved in circles, whose centers themselves moved in other circles, and so on and so forth.

The Copernican idea could, at one stroke, resolve all the discrepancies mentioned above. The only key issue in which Copernicus goofed up was in sticking to the circular motion. As a result, he also needed epicycles for his model to work. Historians do not seem to agree on the actual number of epicycles used by different people. As a measure of complexity, the number of epicycles used was given as 80 for Ptolemy, and a mere 34 for Copernicus in some works [5]. The largest number of epicycles that were mentioned was in the Encyclopedia Britannica on Astronomy during the 1960s, while discussing King Alfonso X of Castile's interest in astronomy. Each planet had been thought of as requiring from 40 to 60 epicycles to represent its orbit. Amazed at the complexity, Alfonso is supposed to have remarked that, had he been present at the creation, he might have given excellent advice [6]. A major difficulty with this epicycles-on-epicycles counting is that historians examining books on Ptolemaic astronomy from the Middle Ages and the Renaissance have found no clear trace of multiple epicycles being used for each planet. And this feature remained until Kepler later changed the circular orbits to elliptical orbits, laying to rest the last of the Aristotelian dogmas.

The book was eminently unreadable and sold poorly. While several other contemporary books on planetary theory and astronomy were easily reaching their 100th reprints in Germany, Copernicus's book stopped after just one printing. All the same, one cannot deny that it was a milestone in science. It was immediately adopted by those who produced the planetary tables and helped to set the heavens in order. What was probably more important, it influenced at least a handful of later thinkers to step away from religious dogma and think anew. It was this rebirth of free thinking which, eventually, heralded the Scientific Revolution in Europe.

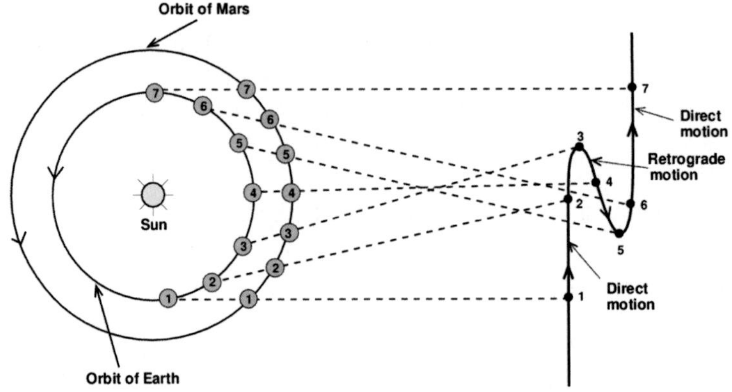

Fig. 11.4: The origin of the 'retrograde' motion of some planets, illustrated using Mars as an example. Since the Earth revolves faster around the Sun than Mars, at times the Earth overtakes Mars in orbital motion. Seen from the Earth, Mars will then appear to move 'backwards' in the sky relative to the Earth.

Notes, References, and Credits

Notes and References

1. Whitehead, Alfred North (1925), *Science And The Modern World*, The New American Library, USA [ASIN B0006AJLTA].
2. For more on Copernicus and his times, see, e.g.,
   Koestler, Arthur (1968), *The Sleepwalkers*, Macmillan, UK [ISBN 978-0140209723].
   Armitage, Angus (1990), *Copernicus, the founder of modern astronomy*, Dorset Press, UK [ISBN 978-0-88029-553-6].
   Bienkowska, Barbara (1973), *The Scientific World of Copernicus: On the Occasion of the 500th Anniversary of His Birth, 1473–1973*, Springer, Netherlands [ISBN 978-94-010-2616-1].
   Finocchiaro, Maurice A. (2010), *Defending Copernicus and Galileo: Critical Reasoning in the Two Affairs*, Springer Science and Business Media, Germany [ISBN 978-9048132003].
   Sobel, Dava (2011), *A More Perfect Heaven: How Copernicus Revolutionized the Cosmos*, Walker & Company, New York [ISBN 978-0-8027-1793-1]. (This book features a fictional play about Rheticus's visit to Copernicus, sandwiched between chapters about the visit's prehistory and post-history.)

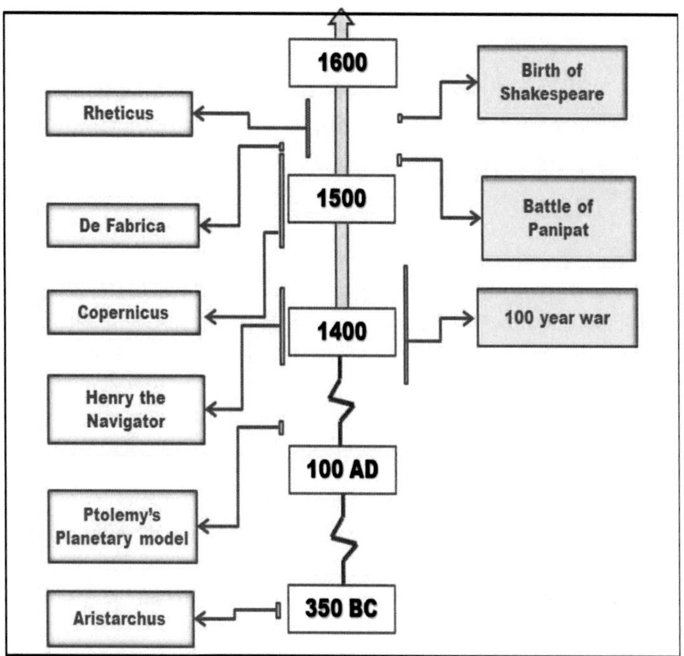

WHEN

3. For more on Aristarchus of Samos, as he was called, see:
Gomez, A. G. (2013), *Aristarchos of Samos, the Polymath*, Author House, USA [ISBN 9781496994233].
Stahl, William (1970), *Aristarchus of Samos in the Dictionary of Scientific Biography*, Charles Scribner and Sons, New York [ISBN 0-684-10114-9].
Heath, Sir Thomas (2013), *Aristarchus of Samos, the ancient Copernicus; a history of Greek astronomy to Aristarchus* (together with Aristarchus's Treatise on the sizes and distances of the Sun and Moon: Greek text with translation and notes), Cambridge University Press reissue edition, Cambridge [ISBN 978-1108062336].
4. For the role played by Rheticus, see, e.g.,
Danielson, Dennis (2006), *The First Copernican: Georg Joachim Rheticus and the Rise of the Copernican Revolution*, Walker and Company, New York [ISBN 0-8027-1530-3].
Westman, Robert (2011), *The Copernican Question: Prognostication, Skepticism, and Celestial Order*, University of California Press, Berkeley [ISBN 978-0520254817].
5. Palter, Robert (1970), *An Approach to the History of Early Astronomy*, Studies in History and Philosophy of Science Part A **1**(2), 93

6. This appears in *Encyclopedia Britannica*, 1968 edition, vol. 2, p. 645. This is identified as the highest number by Owen Gingerich, who also expressed some doubt about the quotation attributed to Alfonso. In his book, *The Book Nobody Read: Chasing the Revolutions of Nicolaus Copernicus* (p. 56, Penguin, London), Gingerich states that he challenged Encyclopedia Britannica about the number of epicycles. Their response was that the original author of the entry had died and its source couldn't be verified!

Figure Credits

7. Figure 11.1 courtesy: From Edward Grant, "Celestial Orbs in the Latin Middle Ages", Isis, Vol. 78, No. 2. (June 1987), pp. 152–173. Wikimedia Commons / Public Domain. https://commons.wikimedia.org/wiki/File:Ptolemaicsystem-small.png (from public domain).
8. Figure 11.2 courtesy: Unknown artist/ Wikimedia Commons/Public Domain. https://commons.wikimedia.org/wiki/File:Nikolaus_Kopernikus.jpg (from public Domain).
9. Figure 11.3 courtesy: Thomas Digges (1546-1595) / Wikimedia Commons/Public Domain. https://commons.wikimedia.org/wiki/File:ThomasDiggesmap.JPG (from public domain).

12

The Logarithm — An Unsung Hero

"My Lord, I have undertaken this long journey purposely to see your person, and to learn by what engine of wit or ingenuity you came first to think of this most excellent help in astronomy" [1]. This was how the Scottish nobleman Henry Briggs (1561–1630) is supposed to have greeted John Napier (1550–1617), the inventor of logarithms, when they met first. And indeed it was an invention of great practical utility, not only in astronomy but also in other branches of science which required extensive computation. Until the advent of electronic calculating devices, logarithms remained the most effective tool for every scientist — and yet, not too many people appreciate this fact.

The inventor of this canny tool, John Napier, was a rather enigmatic personality [2]. He was born into the Scottish aristocracy and travelled widely all over Europe during his youth. This was the time when Europe was in total turmoil, split into different warring camps by the Protestant reformation. His native country, Scotland, was fast turning into a Calvinist state. This influenced Napier and he became a vocal Protestant gentleman and wrote a book, *A Plaine Discovery of the Whole Revelation of Saint John*, which was a tirade against the Church of Rome. Given the popularity the book received, Napier strongly believed that his place in history would be associated with his religious views. Fortunately, he also did more useful things by which we now remember him.

T. Padmanabhan and V. Padmanabhan, *The Dawn of Science*,
https://doi.org/10.1007/978-3-030-17509-2_12

Box 12.1: Another Side of John Napier

In 1593, Napier expressed his anti-Catholic views against the Church of Rome through a book entitled *A Plaine Discovery of the Whole Revelation of Saint John*. In this book, he bitterly attacked the Catholic church, claiming that the Pope was indeed the Anti-Christ and urging the Scottish King James VI (later to become King James I of England) to purge his court of all "Papists, Atheists, and Newtrals". Interestingly, the book also predicted that the Day of Judgement would fall sometime between 1688 and 1700!

The book became an overnight best-seller, was translated into several languages, and ran 21 editions of which at least 10 were during Napier's lifetime! So engrossed was he in his religious sentiments that he definitely expected his claim to fame in history — or what was left of it with the Day of Judgement being rather imminent — to rest on this book.

Around this time, there was an attempt by the Scottish Catholic Earls to gain support from the Spanish King, against Protestantism in Scotland. Napier spent a considerable amount of time designing different kinds of ingenious war machines to be used against Philip II of Spain, in case he should attack Scotland [3]! A document dated 7 June 1596 and carrying the title 'Secret Inventions, profitable and necessary in these Days for Defence of this Island, and withstanding of Strangers, Enemies of God's Truth and Religion' — which was passed on to the English government in July 1596 — described the design of four weapons which Napier had been investigating.

The second Spanish Armada never set sail for British shores, and John Napier's weaponry remained as drawings and descriptions which, upon his instructions, were destroyed a short time after his death.

Given the time he spent on religious matters, it is rather surprising that he also had the time and energy to ponder over mathematics. He was particularly concerned with the amount of labour involved in the multiplication and division of numbers. In fact, most scientists of his time spent a large part of their working day doing routine, boring calculations. This was especially true of the astronomical computations involved in preparing planetary tables, etc. The invention of logarithms changed the situation completely; it replaced multiplication by addition and division by subtraction. As Laplace (1749–1827) said years later, the logarithms effectively 'doubled the lifetime of the astronomer'.

Fig. 12.1: Portrait of John Napier, the inventor of the logarithm [5]. The use of logarithms replaced multiplication by addition and division by subtraction! This saved a huge amount of time and effort for the scientific community. In particular, astronomers of that period used to spend a large part of their time doing the routine, boring calculations which were needed to estimate, say, the planetary orbits. Logarithms came as a great relief to them, and made Laplace comment that the invention of logarithms had effectively 'doubled the lifetime of the astronomer'!

As usually happens with great inventions, the idea behind the logarithm is extremely simple. To understand the basic concept, consider the numbers: $2^0 = 1, 2^1 = 2, 2^2 = 4, 2^3 = 8, 2^4 = 16, 2^5 = 32$, and so on. Suppose we want to multiply the numbers 4 and 8. Since 4 is 2^2 and 8 is 2^3, the product can be written as $4 \times 8 = 2^{2+3} = 2^5$. We know that 2^5 is 32, and this immediately gives us the answer. The crucial point is that the *multiplication* of two numbers 4 and 8 has been reduced to the addition of the superscripts 2 and 3.

Now suppose we have a look-up table of all numbers expressed as various powers of 2. For example, the number 17 can be expressed as $2^{4.087}$ to a very good accuracy and 19 can be written as $2^{4.248}$. (We will say that 4.087 is the 'logarithm' of 17 to the 'base' 2.) Then, to multiply 17 and 19, we only have to add the two powers (4.087 + 4.248) and obtain 8.335. The product 19×17 will be $2^{8.335}$, which is 323. Of course, for this idea to work we need a detailed table expressing all integers as powers of 2. Once such a table is prepared, any two numbers can be multiplied — or for that matter divided, which requires subtracting one index from the other — with ease. This saves a considerable amount of time and effort.

In the above illustration, we used the number 2 as the 'base' to express all other integers. One could have used any other positive number in place of 2. For a rather complicated mathematical reason, Napier used a number which is the reciprocal of a constant, usually denoted by the letter $e \approx 2.718$. This constant plays a crucial role in all branches of higher mathematics and is, in fact, called 'the base of the natural logarithm' [4].

n	$[1+(1/n)]^n$
10	2.593742
100	2.704814
1000	2.716924
10000	2.718146

Fig. 12.2: An interpretation of the number e, used as the base of the natural logarithm, in the context of the calculation of compound interest. If a bank gives you 100 per cent interest (per annum) but computed several times within a year, then your money will grow significantly — but not without limit. Even if the bank computes the interest on an instantaneous basis, your money at the end of the year will only be e times your original investment.

One way to understand this number e is as follows. Suppose you invest 1 euro with a bank which is willing to offer an incredibly high interest of 100 per cent per annum. (Nobody is going to give this much interest, but it makes our calculations simple!) Then, at the end of a year your money will have doubled. That is, it will have gone up by a factor 2. Suppose the bank decides to help you out more by calculating the compound interest on a half-yearly basis, i.e., once every 6 months. Then at the end of 6 months, you will have $[1 + (1/2)]$ euros which will grow to $[1 + (1/2)] \times [1 + (1/2)] = [1 + (1/2)]^2$ at the end of one year. So your investment will have grown to $[1 + (1/2)]^2 = 2.25$ at the end of the year. Now imagine the bank calculating the interest every 4 months. Then you will get $[1 + (1/3)]^3 = 2.37$. And if the bank calculates interest at every third month, you will have got $[1 + (1/4)]^4 = 2.44$. If the interest calculation is done on a monthly basis, you will get $[1 + (1/12)]^{12} = 2.61$, while weekly basis will fetch $[1 + (1/52)]^{52} = 2.69$, and a daily basis will get you $[1 + (1/365)]^{365} = 2.71$ (see Fig. 12.2).

The pattern should be clear now. If the year is divided into n equal parts and the compound interest is computed at the end of each of these n parts, then you will get, at the end of the year, an amount $[1 + (1/n)]^n$. You will find that, as you increase n to larger and larger values, this particular quantity increases but approaches a specific number (see Fig. 12.2). When you take n to be arbitrarily large, the resulting number is denoted by e, which is considered to be the base of the natural logarithm. As we said before, this number has an approximate value of $e \approx 2.718\,268$, correct to six decimal places.

Napier published his discussion in 1614 in a small brochure titled, *Mirifici Logarithmorum Canonis Descriptio* (A Description of the Wonderful Law of Logarithms), which also contained a table giving the logarithms of the sines of the angles for successive minutes of arc. (This was the most important quantity required for astronomical computations.) The book created wide interest. It is not often in the history of science that a new idea gets such an enthusiastic reception (see Fig. 12.3). Napier's invention was quickly adopted by scientists, not only in Europe but also in far-off China. Amongst those to make immediate use of logarithms was Johannes Kepler (1571–1630), who adopted them for his elaborate calculations of the planetary orbits.

The news travelled fast and just a year later, Henry Briggs, a professor of mathematics in London, travelled all the way to Edinburgh (it was quite a distance to travel in those days!) to greet Napier. At their meeting, Briggs convinced Napier that it would be much more useful to use 10 as the base for logarithms instead of $1/e$. In other words, all the numbers should be expressed as powers of 10. Since $100 = 10^2$ and $1000 = 10^3$, the logarithm of 100 in base 10 will be 2 and that of 1000 will be 3; the logarithm of any number between 100 and 1000 will be a number between 2 and 3.

Briggs himself started the construction of such a table on his return to London and, in 1624, he published his *Arithmetica Logarithmica*, containing tables of logarithms (up to the 14th decimal place!) for all the numbers from 1 to 20 000 and from 90 000 to 100 000. The gap between 20 000 and 90 000 was filled in later by a Dutch bookseller, Adrian Vlacq (1600–1666). These tables remained in use for nearly three centuries and were superseded only around the 1940s when extensive 20-decimal place tables were calculated.

Meanwhile, in the 1620s, the English mathematician, William Oughtred (1574–1660), realized that even the process of looking up logarithmic tables can be eliminated by constructing a simple mechanical device. It consisted of two sliding scales in which the numbers are marked in such a way that the their distance from the left end of the scale is numerically equal to the logarithm of the number.

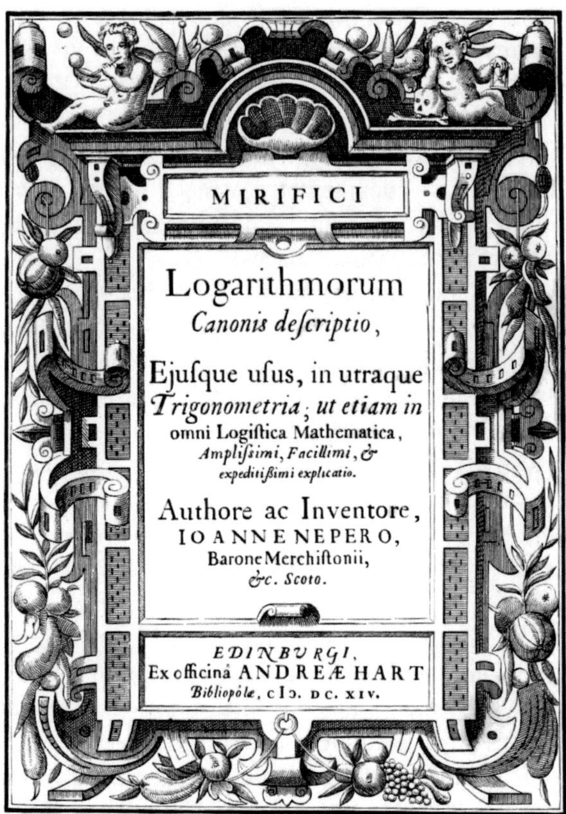

MIRIFICI

Logarithmorum

Canonis deſcriptio,

Ejuſque uſus, in utraque

Trigonometria; ut etiam in

omni Logiſtica Mathematica,

Ampliſsimi, Facillimi, &

expeditiſsimi explicatio.

Authore ac Inventore,

IOANNE NEPERO,

Barone Merchiſtonii,

&c. Scoto.

EDINBURGI,

Ex officinâ ANDREÆ HART

Bibliopôlæ, CIↃ. DC. XIV.

Fig. 12.3: Title page of Napier's book *Mirifici Logarithmorum Canonis Descriptio* (A Description of the Wonderful Law of Logarithms) [6], published in 1614. Napier had spent over twenty years devising his theory of logarithms. In this book, Napier established his technique for working out logarithms and also obtained the logarithms for sines, tangents, and secants of arc length — quantities which were heavily used in astronomical calculations. The book created considerable interest in the scientific community, and its popularity spread far and wide.

Numbers could now be multiplied and divided by merely sliding one scale on another. It is difficult to estimate how much the world of engineering and science owes to this simple gadget called the 'slide rule' (see Fig. 12.4).

Napier also made some contributions to other branches of mathematics – for instance, perfecting the decimal notation which we now use so frequently. The idea of the decimal fraction itself had been worked out earlier by the Dutch mathematician, Simon Stevin (1548–1620), but it was Napier who made the notation compact and convenient to use.

Incidentally, there has been some evidence that Jost Burgi (also known as Byrgius) (1552–1632), a Swiss watchmaker, had also developed the idea of logarithms as early as 1588, six years before Napier began work on the same idea. Burgi constructed a table of — what are now understood to be — anti-logarithms, using a method distinct from Napier's. But for some reason he did not publish it until 1620, when his table was issued anonymously in Prague. Kepler, in the introduction to his Rudolphine Tables, commented rather negatively about Burgi's logarithms: "... as aids to calculation Justus Byrgius was led to these very logarithms many years before Napier's system appeared; but being an indolent man, and very uncommunicative, instead of rearing up his child for the public benefit he deserted it at birth." Today his name is totally unknown, except among historians of science.

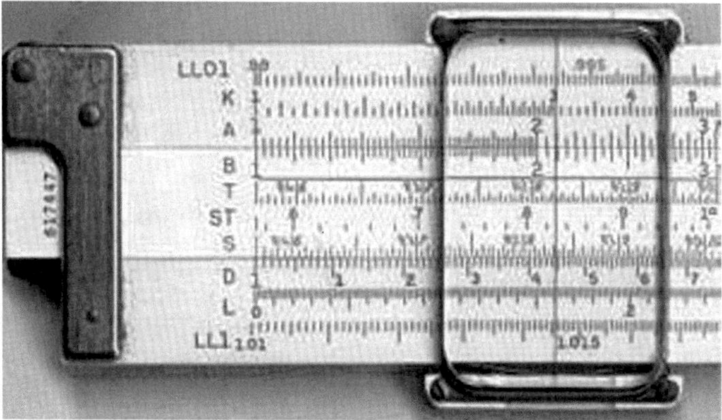

Fig. 12.4: The slide rule, constructed using the properties of the logarithm, was the friend of every engineer and scientist until calculators made it obsolete! [7]

Box 12.2: A Humble Servant of Science

Incredible though it may seem, there *was* a time when we did not have electronic computing devices! A direct off-shoot of logarithms, a device called the slide rule, helped engineers to do the complicated calculations required in their work — from figuring out the double helix to building a Boeing aircraft!

The first design of a slide rule was by William Oughtred, around 1622, in which he used two sliding scales with graduations marked by logarithms, thus allowing multiplication and division by sliding one scale relative to the other. To this, Isaac Newton added the refinement of a glass cursor in 1675, thereby completing the standard design used for decades. A variant of the design, created by Thomas Everard in 1683, called the gauging rule, was even used to determine the content of ale, wine, and spirits barrels so that their excise tax could be calculated properly!

While elementary in its principle, the slide rule underwent several rounds of modifications, and many different special purpose slide rules were designed. Between 1625 and 1800, during the first 175 years after its invention, nearly 40 different kinds of slide rules came into being, including circular and spiral ones. In the next hundred years, the world saw another 250 variations on the slide rule. Historians of technology estimate that about 40 million slide rules were manufactured in the twentieth century alone.

Being a mechanical device and hence immune to electric failure or battery depletion, the slide rule is, even today, kept as a backup by many sailors for navigation, especially on segments of long routes! The slide rule remains in the background of the work which went into everything from the construction of the Empire State building to the Moon landing.

All this is essentially a testimony to the power of logarithms.

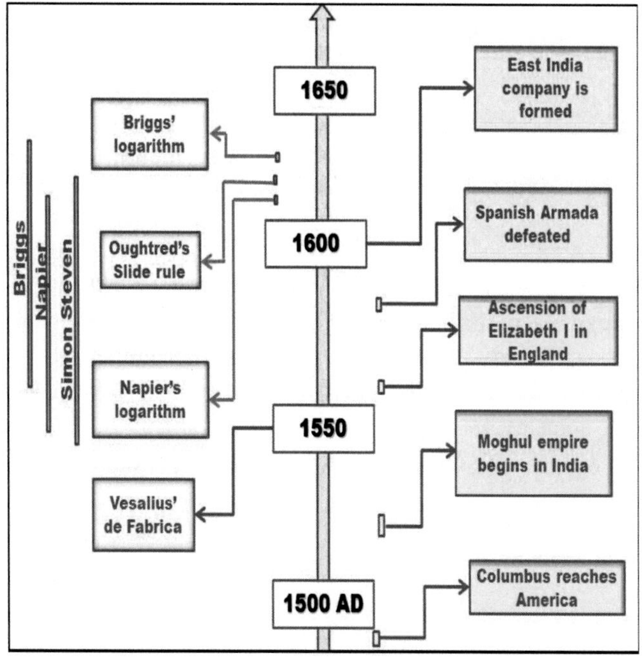

WHEN

Notes, References, and Credits

Notes and References

1. One source for the quote appearing in the first paragraph is: https://www.futilitycloset.com/2014/09/11/likewise/
2. For more on John Napier, see, e.g.,
   Alexandros, Diploudis (1997), *Undusting Napier's Bones*, Heriot-Watt University, UK.
   Cajori, Florian (1991), *A History of Mathematics*, American Mathematical Society, USA [ISBN 978-0-8218-2102-2].
   Bruce, I. (2002), *The Agony and the Ecstasy: The Development of Logarithms by Henry Briggs*, The Mathematical Gazette. **86** (506): 216-227 [doi:10.2307/3621843]
3. See, e.g.,
   Havil, Julian (2014), *John Napier: Life, Logarithms, and Legacy*, Princeton University Press, Princeton [ISBN 978-0691155708].

Seath, Gary (2017), *Beyond Logarithms and Bones*, Chap. 8; a special collection published by Edinburgh Napier University (UK) to commemorate 400 years since his death.

4. For a popular account of e, the base of natural logarithms, see:
   Maor, Eli (1994), *e — The Story of a Number*, Princeton University Press, Princeton [ISBN 978-0691033907].

Figure Credits

5. Figure 12.1 courtesy: Unknown artist / Wikimedia Commons/Public domain. https://commons.wikimedia.org/wiki/File:John_Napier.jpg (from public domain).
6. Figure 12.3 courtesy: Napier, Mark, William Blackwood / Wikimedia Commons/Public domain. https://commons.wikimedia.org/wiki/File:Logarithms_book_Napier.jpg (from public domain)
7. Figure 12.4 courtesy: https://en.wikipedia.org/wiki/File:Slide_rule_scales_front.jpg (from public domain).

13 The Way of the Wanderers

Otte Brahe (1518–1571) was a nobleman and governor of the Helsingborg castle in Denmark who, around 1540, promised his brother, Joergen, that if he had a son the latter could adopt him. But, in 1546, when Brahe did beget a son, he went back on his promise. Joergen was nice enough to wait till a second son was born to Brahe, after which he promptly kidnapped — and adopted — the first son [1].

The son was Tycho Brahe (1546–1601) and he grew up to become the most accurate observational astronomer before the days of the telescope [2]. Tycho's foster-father died when Tycho was still very young, leaving him with a large inheritance. (Joergen's death was also rather unusual; he jumped into a river to rescue Ferdinand II, the King of Denmark; though he succeeded in the attempt, he caught pneumonia and died.)

In 1566, Tycho Brahe went to study medicine at the University of Rostock. While there, in December 1566, Tycho ended up losing part of his nose in a sword duel against a fellow Danish nobleman, Manderup Parsberg. Tycho had quarrelled with Manderup on two earlier occasions and they decided to resolve the issue — historians are unclear what it was about — with a duel. The duel (in the dark) resulted in Tycho losing the bridge of his nose and getting a broad scar across his forehead, but the two rivals were reconciled after the incident. Tycho received the best possible medical care; but for the rest of his life he had to wear a prosthetic nose, supposed to have been made of silver and gold, kept in place with glue. (Incidentally, in November 2012, Danish and Czech researchers did a chemical

T. Padmanabhan and V. Padmanabhan, *The Dawn of Science*,
https://doi.org/10.1007/978-3-030-17509-2_13

analysis of a small bone sample from the nose of the body exhumed in 2010 and found that that the prosthetic was actually made out of brass!)

Though he lost a nose, he gained an excellent education and became enamoured of astronomy. Right from his student days, he had kept a careful record of the night sky, day after day. The crucial turning point in his life came in August 1563 when he was observing the conjunction — viz., the close alignment — of Jupiter and Saturn. To his dismay, he discovered that all the almanacs used in those days were widely off the mark in predicting this event! This convinced him of the need for more accurate observations with good instruments — a task he set for himself. He travelled all over Europe, acquiring the necessary instruments, and set up a small observatory at Scania in 1571.

He had the chance, quite literally, of a lifetime, on 11 November 1572, when he spotted a 'new star' near the constellation of Cassiopeia. This new star was brighter than Jupiter. Around 16 November 1572, it reached its peak brightness and remained visible to the naked eye into early 1574, gradually fading out of view. Tycho observed that the new object showed no daily motion (called parallax) with respect to the background of the fixed stars, so it was at least as far away as the Moon and those planets that did show such parallaxes. He also found that the object did not move relative to the fixed stars over the time scale of several months, while all planets (including the outer planets, for which no daily parallax was detectable) did because of their orbital motions. This suggested that it was not even a planet, but a fixed star in the stellar sphere beyond all the planets. In short, Tycho's careful observations showed that this luminous object was a very long way away from the Earth and therefore lay among the fixed stars.

This conclusion, spelt out in Tycho Brahe's book, *De Nova Stella* in 1573 (see the left-hand image in Fig. 13.2), shattered the prevailing dogmas. For, according to the accepted Aristotelian principles, all changes and decay were confined to the Earth, and the realms of the stars were thought to be immutable. The appearance of a new star was therefore a serious blow to this idea.

We now know that it was a supernova, arising from a huge explosion that occurs at the end stages in the evolution of massive stars. Supernovas are among the brightest events in the sky, and this particular one in the constellation of Cassiopeia is one of about eight supernovas which were visible to the naked eye in historical records. It appeared in early November 1572 and was independently noticed on several continents. For example, in England, Queen Elizabeth asked the mathematician and astrologer Thomas Allen (1542–1632), "to have his advice about the new Star that appeared in the Cassiopeia to which he gave his Judgement very learnedly", as we learn from the record kept by John Aubrey (1626–1697) in

Fig. 13.1: Portrait of Tycho Brahe (1546–1601), the astronomer who produced the most accurate observations of the skies, before the invention of the telescope [8]. His book, *De Nova Stella*, published in 1573, described the 'new star' — which we now know was a supernova (see Box 13.1). It was located near the constellation of Cassiopeia and he found it to be brighter than Jupiter and much farther away, in the abode of stars. This shattered the prevailing myth that nothing changes in the heavens and brought Tycho the royal patronage of Denmark's king, Frederick II. The king gave him the island of Ven and the necessary financial support to build an astronomical observatory, which Tycho used very well.

his memoranda a century later. In China, which was ruled by the Ming dynasty, the star was interpreted as an evil omen and the emperor was warned to behave with propriety!

Box 13.1: Supernovas: The Paradigm Breakers

Massive stars end their life in a glorious explosion — called a supernova — which constitutes one of the most fascinating optical displays in the sky. A supernova in our galaxy can be extremely bright, sometimes brighter than Venus and in some cases visible during the daytime. They appear unexpectedly; even today, while we understand them reasonably well, we cannot *predict* them! They bloom into prominence and then fade away over a period of several weeks. They occur in our galaxy somewhat rarely, at the rate of about two per century.

A total of about 8 supernovas have been observed in our galaxy by the naked eye. The first written evidence of a supernova — seen in retrospect — was as early as 185 AD, when Chinese astronomers spotted a 'new star' in the night sky, between the constellations of Circinus and Centaurus, which shone for eight months. Astronomers now believe it was probably the supernova whose remnant is now catalogued as SN 185.

Another supernova, SN 1006, was probably the brightest such event in recorded history. Appearing in the constellation of Lupus in 1006 AD, this was more than sixteen times brighter, at its peak, than Venus. Not surprisingly, it was recorded by observers right round the globe — in China, Japan, Iraq, Egypt, and Europe, and possibly also in North American petroglyphs. Yet another famous one was SN 1054, whose remnant, called the Crab nebula, is one of the most extensively studied objects in the sky. First observed on 4 July 1054 by Chinese astronomers, it remained visible for about two years. Finally, it is amusing to note that Nature accorded a supernova, not only to Tycho but also to Kepler. Kepler discovered SN 1604, in the constellation of Ophiuchus, just a few decades after Tycho's observation — which is really a rare coincidence.

Tycho's supernova, as we have seen, shattered the Aristotelian paradigm that the heavens are perfect and unchanging. Nearly 400 years later, the observations of distant supernovas by the Hubble telescope led to another major paradigm shift: astronomers discovered [3] that, not only is our universe expanding, it is also speeding up, i.e., the expansion rate itself is increasing! Such an accelerated expansion of the universe demands the existence of what cosmologists call 'dark energy'. The nature of dark energy remains a major theoretical puzzle today [4].

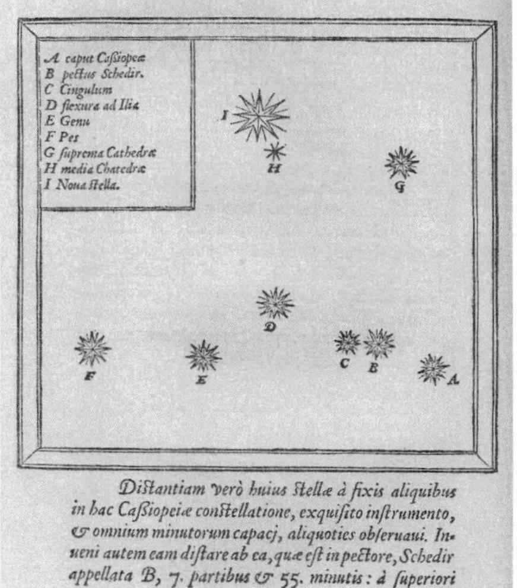

A caput Caßiopeæ
B pectus Schedir.
C Cingulum
D flexura ad Ilia
E Genu
F Pes
G ſuprema Cathedræ
H media Chatedræ
I Noua ſtella.

I H G D F E C B A

Diſtantiam verò huius ſtellæ à fixis aliquibus in hac Caßiopeiæ conſtellatione, exquiſito inſtrumento, & omnium minutorum capacj, aliquoties obſeruaui. Inueni autem eam diſtare ab ea, quæ eſt in pectore, Schedir appellata B, 7. partibus & 55. minutis: à ſuperiori verò

Fig. 13.2: Left: The star map from Tycho Brahe's *De Nova Stella*, showing the position of the supernova of 1572 in the constellation of Cassiopeia (labelled I) [9]. The objects marked A to H are known stars which belong to the same constellation. Tycho described this supernova of 1572 (which is now known as "Tycho's Supernova" or SN 1572) in his book *De Nova Stella*. This work contained both Tycho Brahe's own observations and also the analysis from many other observers. Though he may not have been the first to observe the 1572 supernova, he was probably the most accurate observer of the object [5]. This supernova observation also challenged the Aristotelian dogma of the unchangeability of the realm of stars. Right: A recent (2010) image from NASA's Wide-field Infra-red Survey Explorer (WISE) mission, showing the region containing the supernova remnant. The red circle in the upper left part of the image indicates SN 1572, "Tycho's Supernova" [10].

Fig. 13.3: Tycho Brahe's observatory on the island of Ven was built with a very high level of workmanship, which helped in making very accurate observations of heavenly bodies [11]. Tycho improved on virtually every single astronomical measurement available at his time. For example, he observed the motions of the planets in the sky, especially that of Mars, with very high accuracy. This was important for Kepler — who used Tycho's data — to discover the correct laws of planetary motion and to distinguish between competing models. His determination of the length of the year, to an accuracy of almost a second, was useful later on in the calendar reforms (see Chap. 21).

This discovery also brought Tycho royal patronage. Denmark's king, Frederick II, gave him the island of Ven and also the necessary financial support to build an astronomical observatory. Tycho used it efficiently; the observatory (Fig. 13.3) was a masterpiece of workmanship and he was extremely accurate in making observations. To put it in perspective, Ptolemy's observations were correct to 10 minutes of arc, while Tycho's were exact up to two minutes of arc!

In 1577, a great comet appeared in the sky, and Tycho kept a careful track of its path. His measurements again confirmed two facts: (i) the comet was at a much greater distance than the Moon, and (ii) its path was significantly different from a circle. These results further damaged the credibility of Aristotle's ideas.

Tycho corrected, for the better, every single astronomical measurement which existed in his day. For example, he observed the motions of the planets, especially that of Mars, with unprecedented accuracy. He also determined the length of the

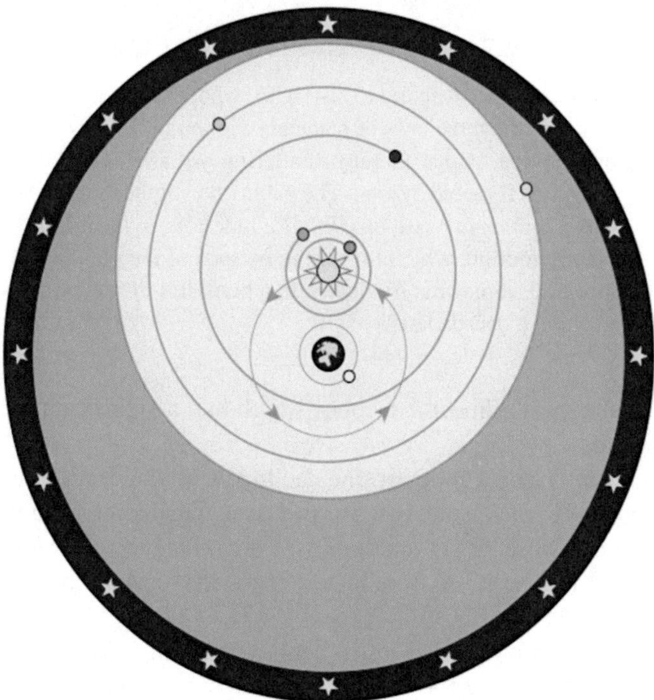

Fig. 13.4: Tycho Brahe's model of the Solar System in which the Sun revolved around the Earth, but the other planets revolved around the Sun [12].

Fig. 13.5: Portrait of Johannes Kepler (1571–1630), who used Tycho's extremely accurate observations to discover the three laws of planetary motion [13]. In particular, he replaced circular paths by elliptical paths, thereby shattering yet another myth of the heavens reflecting the symmetries favoured by men. The publication of the first two laws of planetary motion, in his book *Astronomia Nova*, in 1609, is a milestone in the history of science. The third law of planetary motion, which linked the periods of the planets to their distances from the Sun, appeared, somewhat ironically, in a book full of useless mysticism, which Kepler published about a decade later!

year to an accuracy of almost a second, which had a significant bearing on the calendar reforms.

Unfortunately, Tycho's good fortune declined with the death of Frederick II in 1588. He managed to pick fights with the new king, the nobility, and the clergy, and had to abandon his observatory and finally settle in Prague, under the patronage of Emperor Rudolf II. And it was here that he made his most important discovery — Johannes Kepler (1571–1630).

It was a study in contrast, Kepler being everything Tycho was not [6]. He was born in Germany to a good-for-nothing mercenary soldier and a quarrelsome mother who, in her later years, almost got herself burnt at the stake as a witch [7]; he was sickly and depressed, but managed a good education only because his

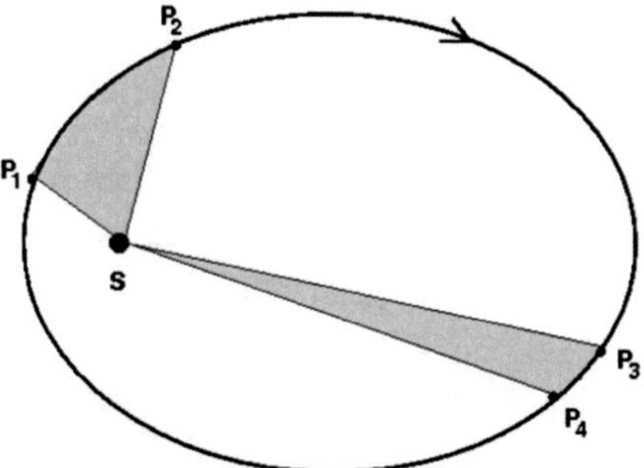

Fig. 13.6: Illustration of Kepler's second law. The blue regions have equal areas and the planet takes the same amount of time to go from P_1 to P_2 as to go from P_3 to P_4.

superior intelligence was fortunately recognized by the local duke, who gave him a scholarship.

Kepler was training to become a Lutheran minister after his formal education at Tubingen University but, purely by accident, became a mathematics teacher in a Lutheran school in Graz in Austria. (He was recommended for this job by the university, on the death of the former teacher.) As a student of astronomy, Kepler was strongly influenced by the Copernican concepts. His early attempts were to connect the planetary orbits with the Platonic solids (Box 2.2), which were not very successful; but they brought him into contact with Tycho. In 1597, as religious disputes became intense in his home town, Kepler moved to Prague and accepted a position in Tycho's observatory. With the death of Tycho in 1601, Kepler inherited the vast amount of astronomical observations which Tycho had maintained over the years.

Kepler started, though rather erratically, to find simple rules describing the motion of the planets — especially that of Mars. The failure of the simpler models (using circles, see Fig. 13.4) finally forced Kepler to the conclusion that "the path of planets around the Sun are ellipses with the Sun at one focus". The extraordinary accuracy of Tycho's observations was instrumental in eliminating several other simpler models; some of these models differed in their predicted positions, from those based on elliptical orbits, by just eight minutes of arc — which would have been irrelevant in Ptolemy's day.

Though Kepler replaced the circular paths by a less symmetrical elliptical path, he was able to discover another symmetry in the motion of the planets. Using once again Tycho's meticulous observations, Kepler could conclude that "the line joining the planet and the Sun traverses equal areas in equal amounts of time" (Fig. 13.6). These two laws, published in his book *Astronomia Nova* (Fig. 13.7) in 1609, deservedly earn Kepler a place in the history of science. A decade later, he published another book, full of mysticism, in the middle of which lies a gem relating the period of revolution of the planet to its distance from the Sun: "The square of the period of revolution of a planet is proportional to the cube of its distance from the Sun"!

Box 13.2: The Ellipse

The ellipse — the orbits followed by the planets — belongs to a class of curves called 'conics' or 'conic sections', first studied in detail by the Greek geometer Apollonius (third century BC). Apollonius showed that three different kinds of curves can be generated when a plane intersects a cone. The ellipse, in particular, is the only closed curve out of these three (with the circle treated as a special case of an ellipse).

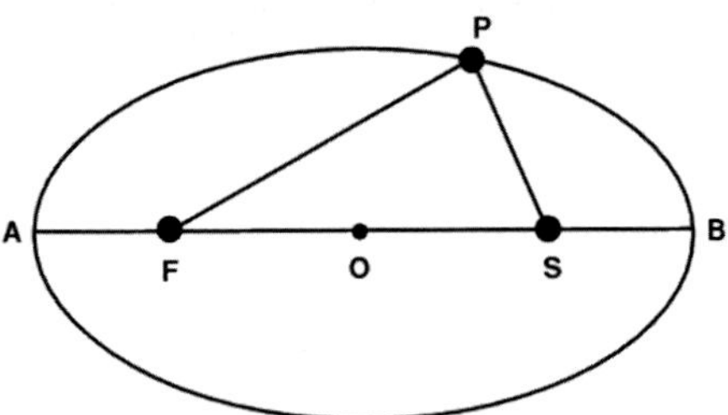

A more physical way of describing the ellipse would be the following. Think of a person walking in such a way that the sum of her distances from two different points F and S (see figure above) remains constant; her path will be an ellipse. The fixed points are called the 'foci' of the ellipse. When these two points coincide, the ellipse becomes a circle; the farther the points, the more elongated the ellipse will be, compared to the circle.

According to Kepler's laws, the planets orbit around the Sun on elliptical paths with the Sun located at one of the foci. Notice that Kepler's second law (Fig. 13.6) says that it takes the same time for the planet to move from P1 to P2 as it does to go from P3 to P4. This requires the planet to move faster while it is nearer the Sun, once again suggesting the influence of the Sun on the planetary orbits.

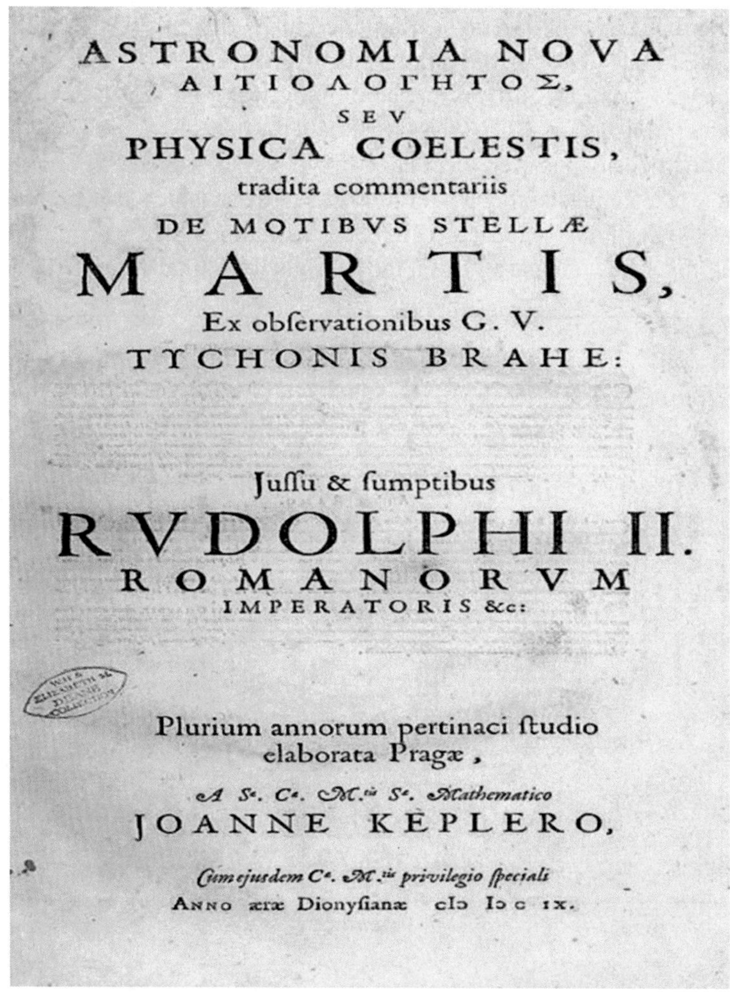

ASTRONOMIA NOVA
ΑΙΤΙΟΛΟΓΗΤΟΣ,
SEV
PHYSICA COELESTIS,
tradita commentariis
DE MOTIBVS STELLÆ
MARTIS,
Ex obſervationibus G. V.
TYCHONIS BRAHE:

Juſſu & ſumptibus
RVDOLPHI II.
ROMANORVM
IMPERATORIS &c:

Plurium annorum pertinaci ſtudio
elaborata Pragæ,
A S^{e}. C^{e}. $M.^{tis}$ S^{e}. *Mathematico*
JOANNE KEPLERO,
Cum ejuſdem C^{e}. $M.^{tis}$ *privilegio ſpeciali*
ANNO æræ Dionyſianæ cIↄ Iↄc ix.

Fig. 13.7: The title page of Kepler's book, *Astronomia Nova* (New Astronomy), published in 1609, which contained the first two of Kepler's three laws of planetary motion [14].

This third law of Kepler's took a significant step in a new direction. By relating the orbital properties of the planets to the central agency, the Sun, it almost immediately suggested that the Sun must be the *cause* of planetary motion. This idea was very much present — rather implicitly — in Kepler's writings, but it took the genius of Isaac Newton to form a workable law out of this suggestion.

During the years 1620 to 1627, Kepler completed the new tables of planetary motions based on Tycho's observations and his own theory of planetary orbits. In spite of severe financial difficulties, continuing war, and religious unrest, these tables were published in 1627. They were called Rudolphine tables, in honour of Kepler's first patron, and were dedicated to the memory of Tycho. Incidentally, it would have taken considerably longer to work out these tables, had Kepler not been able to use logarithms (see Chap. 12)! The Rudolphine tables also contained a set of logarithms. Thus, the pathways in the heavens were finally charted.

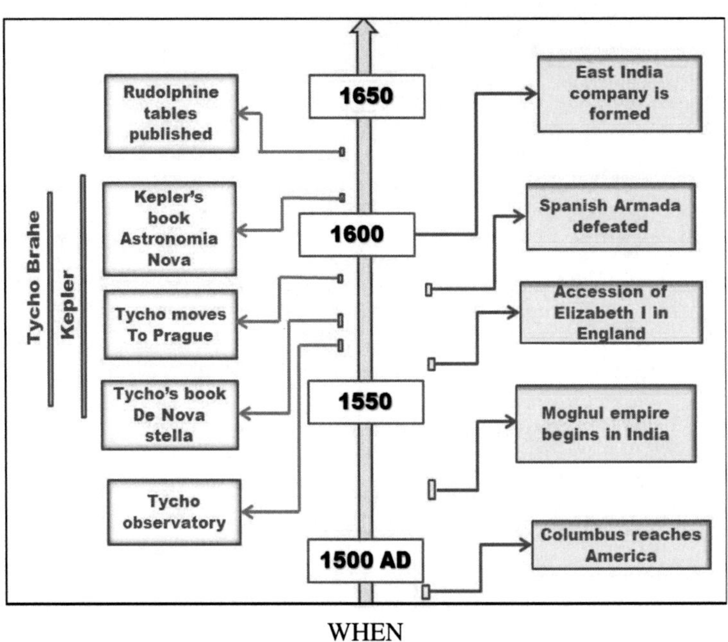

WHEN

Notes, References, and Credits

Notes and References

1. Dreyer, John Louis Emil (1911), *Tycho Brahe: A Picture of Scientific Life and Work in the Sixteenth Century*, Peter Smith Pub Inc, USA [ISBN : 978-0844619965].
2. For more on Tycho Brahe and his times, see:

Christianson, J. R. (1999), *On Tycho's Island: Tycho Brahe and His Assistants 1570–1601*, Cambridge University Press, Cambridge [ISBN 978-0521650816].
Gingerich, Owen (1973), *Copernicus and Tycho*, Scientific American, **173**, 86–101.
Linton, Christopher M. (2004), *From Eudoxus to Einstein — A History of Mathematical Astronomy*, Cambridge University Press, Cambridge [ISBN 978-0-521-82750-8].
Mosley, Adam (2007), *Bearing the heavens: Tycho Brahe and the astronomical community of the late sixteenth century*, Cambridge University Press, Cambridge [ISBN 9780521838665].

3. This discovery was honoured with a Nobel Prize for Saul Perlmutter (USA), Brian P. Schmidt (Australia), and Adam G. Riess (USA) in the year 2011. [https://www.nobelprize.org/prizes/physics/2011/press-release]
4. Padmanabhan T., *The Universe Began with a Big Melt, Not a Big Bang: The cosmological constant and the creation of the universe*, Nautilus, Issue 53, October 5, 2017. [http://nautil.us/issue/53/monsters/the-universe-began-with-a-big-melt-not-a-big-bang].
5. Calar Alto Observatory-CAHA, (December 2008) "Blast From The Past: Astronomers Resurrect 16th-Century Supernova", ScienceDaily. <www.sciencedaily.com/releases/2008/12/081203133809.htm>.
6. For more on Kepler, see:
Koestler, Arthur (1968), *The Sleepwalkers*, Macmillan, UK [ISBN 978-0140209723].
Ball, Sir Robert Stawell (2016), *Great Astronomers: Johannes Kepler*, FeedBooks, USA [ISBN 978-1541166509].
Caspar, Max (1993), *Kepler (Dover Books on Astronomy)*, Dover Publications, USA [ISBN 978-0486676050].
Love, David K. (2015), *Kepler and the Universe: How One Man Revolutionized Astronomy*, Prometheus Books, New York [ISBN 978-1633881068].
7. See, e.g.,
Connor, James A. (2005), *Kepler's Witch: An Astronomer's Discovery of Cosmic Order Amid Religious War, Political Intrigue, and the Heresy Trial of His Mother*, HarperOne, USA [ISBN 978-0060522551].
Rublack, Ulinka (2015), *The Astronomer and the Witch: Johannes Kepler's Fight for his Mother*, Oxford University Press, Oxford [ISBN 978-0198736776].

Figure Credits

8. Figure 13.1 courtesy: By Eduard Ender (1822-1883) Wikimedia Commons / Public domain.
https://commons.wikimedia.org/wiki/File:Tycho_Brahe.JPG (from public domain).
9. Figure 13.2 (left) courtesy: Brahe Tychonis Wikimedia Commons / Public domain.
https://commons.wikimedia.org/wiki/File:Tycho_Cas_SN1572.jpg (from public do-

main).

10. Figure 13.2 (right): Image credit: NASA/JPL- Caltech/UCLA, https://www.jpl.nasa.gov/spaceimages/details.php?id=PIA13119
11. Figure 13.3 courtesy: Unknown / Wikimedia Commons/Public domain. https://commons.wikimedia.org/wiki/File:Tycho-Brahe-Mural-Quadrant.jpg (from public domain).
12. Figure 13.4 courtesy: Fastfission / Wikimedia Commons/Public domain. https://commons.wikimedia.org/wiki/File:Tychonian_system.svg (from public domain).
13. Figure 13.5 courtesy: Unknown/ Wikimedia Commons/Public domain. https://commons.wikimedia.org/wiki/File:Johannes_Kepler_1610.jpg (from public domain).
14. Figure 13.7 courtesy: Johannes Kepler/ Wikimedia Commons/Public domain. https://commons.wikimedia.org/wiki/File:Astronomia_Nova.jpg (from public domain).

14

Galileo and the Dynamical World

Almost around the same time as when Kepler was perfecting the laws of planetary motion, another man was laying the foundations of theoretical mechanics. This was Galileo Galilei (1564–1642) — probably the most misunderstood of all the medieval scientists. Contrary to popular lore, he did not invent the telescope or the thermometer or the pendulum clock; nor did he discover sunspots; and he never dropped the weights down from the tower of Pisa, nor was he tortured by the Inquisition. All these and more are attributed to Galileo from time to time purely because of his historical importance!

Galileo was born on 15 February 1564 in the town of Pisa in Italy [1]. His father wanted him to practice medicine — mainly because doctors used to earn considerably more than mathematicians, even in those days. It was an accidental exposure to a lecture in geometry that made Galileo turn to mathematics and later to physics.

His first invention was a hydrostatic balance, about which he wrote a short essay in 1586. This was followed up by a treatise on the center of gravity of solids, published in 1589. These works won for Galileo the acclaim of the Italian scholars and the sponsorship of the Duke of Tuscany, Ferdinand de Medici and, finally, the position of mathematics lecturer at the University of Pisa. Later on, in 1592, he became professor of mathematics at the University of Padua where he remained for the next 18 years. (Incidentally, when Galileo was a professor at the University of Padua, there was a student named William Harvey (1578–1657), who got his

T. Padmanabhan and V. Padmanabhan, *The Dawn of Science*,
https://doi.org/10.1007/978-3-030-17509-2_14

medical degree there in 1602, and went on to discover the circulation of blood! See Chap. 15.)

Fig. 14.1: Portrait of Galileo (1564–1642), whose contributions place him in an elite class of four physicists, the other three being Archimedes, Newton, and Einstein [6].

It was probably during this time that Galileo made his fundamental contributions to mechanics, relating to the nature and cause of motion (though he published these results much later). In those days, the ideas that prevailed about the motion of bodies were still those of Aristotle. According to these doctrines, some force was assumed to be necessary to keep a body in motion, even at uniform speed. Followers of Aristotle used to argue that an arrow shot from a bow moves only because the air behind the arrow pushes it. It was also believed that a body dropped from some height would fall with a steady speed; that is, it would cover equal distances in equal times. This speed was supposed to be greater for heavier bodies, which implied that heavier bodies should fall faster than lighter ones.

Galileo's investigations — which, incidentally, anyone could have carried out centuries before — proved all these to be incorrect. He realized that a body dropped from a height acquires increasingly greater speeds as it falls. In fact, he found that the distance it covers increases as the square of the time in flight. That is, if the body travels, say, five metres in the first second of its flight, it will cover twenty metres by the end of two seconds, rather than just ten metres.

What was most remarkable about Galileo's discovery was the approach he used to get there. He decided that the proper way to settle any such question was by direct observation! This was quite difficult to achieve in those days, when accurate time-keeping instruments were not available. Galileo got around the problem by an ingenious trick, viz., making bodies roll down a gently inclined plane rather than dropping them vertically. The inclined plane slowed the bodies down considerably and Galileo could time them using his pulse as a clock! (Curiously enough, Galileo *did* know that a swinging pendulum was an instrument that maintains good periodicity. Somehow it never occurred to him to design a clock using the pendulum.) Galileo also realized that all bodies would fall to the ground from a given height in the same time, if air resistance did not affect them.

The fact that a steady force acting on a body continuously increases its speed raises the immediate question: is it even necessary to have a force acting on a body to keep it moving at constant speed? Galileo thought — quite correctly — that motion with constant speed required no external agency to act on the body. This principle — called the principle of inertia, in modern usage — played a crucial role in the later development of dynamics and in the theory of relativity. Galileo also used this principle to calculate the trajectories of projectiles thrown from the ground.

Though these were significant insights, Galileo did not rise to prominence for these studies, but for a very different reason. Around 1608, a spectacle-maker in Holland, Johann Lippershay (1570–1619), had invented an optical tube containing two lenses — one at each end of the tube — which could make distant objects appear closer. Lippershay sold several of these gadgets in the cities of Europe, and Galileo came to know of this invention in the spring of 1609. Knowing the design, Galileo could easily make a telescope for himself (Fig. 14.2) with a magnifying power of about 30 and he turned the new invention towards the sky. Thus Galileo heralded the age of telescopic astronomy.

Fig. 14.2: Replica of the earliest surviving telescope devised by Galileo Galilei, on display at the Griffith Observatory [7].

Using his telescope, Galileo discovered several aspects of nature which, until then, had not been noticeable to the human eye. He found that the Moon had mountains and the Sun had dark spots, once again showing Aristotle to be totally wrong in his assumptions that only the Earth had blemishes in the form of irregularities and distortions.

The stars and the planets appeared very different through the telescope and Galileo could see many more stars than were visible to the naked eye. All these made him conclude that the stars were much farther away than the planets. More dramatically, Galileo found that Jupiter was surrounded by four subsidiary objects, which orbited it regularly. Within a few weeks of observation, he could work out the periods of each of these satellites. Named Io, Europa, Ganymede, and Callisto, these satellites clearly showed that not all celestial objects went around the Earth or the Sun. His telescope also revealed that Venus had phases (like the Moon), and he spotted some ring-like structures around Saturn. (He could not see the rings well enough to figure out their precise nature; that was done, a few decades later, by Huygens and Cassini; see Chap. 19.)

To be sure, there were other astronomers who were exploring the skies with the telescope at the same time as Galileo. Indeed, the first reports on observations of sunspots, for example, came from Father Scheiner, a Jesuit astronomer. Galileo entered into a long and bitter controversy over priorities in these discoveries, thereby making several powerful enemies.

Galileo announced his initial discoveries in a periodical, which he called *Sidereus Nuncius* (The Starry Messenger). These announcements definitely caught the fancy of the public, especially thanks to the author's fine expository skills. Initially, however, his peers in the scientific community were reluctant even to look through the telescope, and were persuaded to come round only after the leading astronomer of the day, Johannes Kepler, threw his weight behind the discoveries.

Box 14.1: Galileo's Anagram

As we have seen, during Galileo's time there were two models of the universe. The geocentric universe of Claudius Ptolemy (of the second century AD) and the heliocentric universe of Nicholas Copernicus (proposed in 1543). In the Ptolemaic system, the motion of Venus was in a circle (epicycle) with the center of this epicycle lying between the Earth and the Sun. In the Copernican model, on the other hand, Venus went around the Sun.

In 1610, one of Galileo's (former) students, Benedetto Castelli, pointed out to him that, in the Ptolemaic model, Venus — being always illuminated from one side — should always appear crescent-shaped when seen from the Earth. On the other hand, in the Copernican model, Venus would have a crescent shaped appearance when it was on the near side of the Sun but would exhibit a nearly full illuminated circle when it was on the far side of the Sun.

In October 1610, Galileo, with the aid of his telescope, observed the gibbous phase of Venus. He realized the importance of this result immediately. He also knew that anyone else with access to a telescope could reach the same conclusion, viz., that the observations favoured the Copernican model. Galileo wanted to claim priority for the discovery and, at the same time, gain some more time for observing Venus.

By December 1610, Venus had waned to nearly half the illuminated shape and Galileo decided to wait no longer. Following the strategy of those days, Galileo 'published' his results in a Latin anagram that he promised to decipher at a later date. Here is the anagram [2]:

Haec immatura a me iam frustra leguntur o.y.

which means, "These are at present too young to be read by me". By the beginning of 1611, Venus had moved around to the near side of the Sun, and was in its crescent phase. Galileo now unscrambled his anagram to read:

Cynthiae figuras aemulatur mater amorum

which translates to "The mother of love imitates the shape of Cynthia". In more transparent form, this means Venus (considered to be the mother of love) manifests all the phases exhibited by the Moon (Cynthia). This was, of course, possible only if Venus passed on both sides of the Sun during its orbit.

In 1611, Galileo visited Rome and was treated with great honour. The Cardinal and Pope Paul VI gave him friendly audiences and the Jesuit Roman College honoured him with various ceremonies which lasted for a whole day. There were astronomers in that college who not only accepted Galileo's discoveries *in toto*, but also improved on his observations, especially on the phases of Venus. At this stage there was no open animosity between the Church and Galileo.

From such a friendly start, just how a parting of ways between science and the Church arose is an interesting tale in the history of science. While it is usual to universally condemn the Church in this matter, a careful study of historical facts

Fig. 14.3: Galileo demonstrated the power of the telescope [8] to the Doge of Venice, Antonio Priuli, and other members of the Venetian Senate on 21 August 1609, and also gifted a telescope to the Senate. From the Campanile di S. Marco (the bell tower of St Mark's Basilica) in Venice, he showed them the capabilities of the instrument, which allowed them to see distant ships nearly two hours before they were visible to the naked eye.

indicates that Galileo's personality actually went a long way toward aggravating the situation. Galileo acted and wrote in a way that made him several enemies, some quite powerful and influential. The personality of Galileo and also the antagonism of his 'scientific' colleagues were as instrumental in bringing about his conflict with the Church as the Church itself (see Box 14.2).

Box 14.2: Galileo, Kepler, and the Church

Galileo's trial has attracted a tremendous amount of attention in the history of science. It is rather curious to see how the events actually unfolded [3].

In 1597, Galileo received a copy of Kepler's book, *Cosmic Mystery*, the preface to which contained detailed arguments in support of the Copernican theory. Galileo sent him a letter saying, "I adopted the teaching of Copernicus many years ago [. . .] I have written many arguments in support of him and in refutation of the opposite view — which I dared not bring into public light [. . .] frightened by the fate of Copernicus [. . .] who [. . .] is, to a multitude, an object of ridicule and derision."

Kepler replied, pleading with Galileo to come out openly in support of the Copernican model. Galileo refrained from doing so, and instead, stopped communicating with Kepler [4]. And for the next 16 years (a period in which the Church did permit several discussions about the Copernican model), Galileo was actually teaching Ptolemy's ideas in his lectures!

During these 16 years, Kepler had repeatedly communicated his findings to Galileo, but Galileo did not respond to any of these; in fact, he continued to ignore Kepler's scientific contributions (especially the elliptical nature of planetary orbits) and continued to use the old Copernican ideas of celestial bodies moving in circles and epicycles. In spite of this, Kepler always treated Galileo with great generosity and wrote openly, in 1610, supporting the discoveries he had made using the telescope. And in fact, this was the time when Galileo badly needed Kepler's support. Even then, Galileo refused to reciprocate the friendly gestures.

In 1611, the head of the Jesuit Roman College, Cardinal Ballarmine, asked the Jesuit astronomers for their official opinion on Galileo's new discoveries. These astronomers — who had, in fact, improved upon the work of Galileo on the phases of Venus — had no hesitation in giving Galileo a clean bill of health and agreeing that at least Venus went around the Sun, which was the simplest explanation for its phases. The world system suggested by many Church astronomers in those days actually had the planets orbiting the Sun with the Sun itself going around the Earth.

But, in the years to follow, Galileo was forced to enter into controversies with jealous colleagues, Church astronomers, powerful members of the nobility, and many others, on whether the motion of the Earth around the Sun could be proved. The Church was willing to accept Copernican ideas purely as mathematical hypotheses but demanded — may be quite correctly — incontrovertible proof if the scriptures were to be reinterpreted. Galileo found himself at a loss in providing the 'proof', not in the least because he did not want to give credit to Kepler for the elliptical orbits; with circular orbits, Copernican models were as bad as those of Ptolemy and the compromise models did much better.

In 1614–1615, Galileo wrote a few open letters in which he supported the Copernican model, suggested the scriptures had to be reinterpreted, and even tried to argue that the burden of proof should rest with the Church. What was more, he went for a direct showdown with the Pope based on what he considered to be a 'proof' of the motion of the Earth — the proof was based on a completely incorrect theory about the origin of tides.

From then on, events took on an ugly turn. The Pope asked the Qualifiers of the Holy Office to take a clear stand on the matter and this they did on 23 February 1616 — categorically against the idea of the Earth moving around the Sun. The Church decreed that: "The doctrine that the Earth is neither the center of the universe nor immovable, but moves, even with a daily rotation, is absurd, and both philosophically and theologically false, and at the least an error of Faith."

Six days later, Galileo had an audience with the Pope and was firmly told not to exceed the limits set by the Church. (What exactly he was told is a matter of another major historical controversy.) The Holy Office put Galileo on trial in 1633, essentially on the charge that the contents of his book, *Dialogue Concerning the Two Chief World Systems* published in 1632, went against the Canonical decree of 1616. Found guilty, he was allowed to spend the rest of his life under house arrest. (Incidentally, for what it is worth, in 1992 the Pope apologised for the treatment meted out to Galileo!)

Over the years from 1611 to 1632, Galileo completed his magnum opus, *Dialogue Concerning the Two Chief World Systems*, in which he had two people, one representing Ptolemy and the other Copernicus, present their arguments before an intelligent layman. Needless to say, Galileo made the Copernican theory come out on top in the arguments.

Galileo died on 8 January 1642 while still under house arrest, as per the verdict of the Church. His bones rest in the Pantheon of the Florentines, the Church of Santa Croce, next to those of Michelangelo, and Machiavelli, occupying an honoured position. His epitaph was written for him by posterity: *Eppur si muove* (nevertheless it moves) [5].

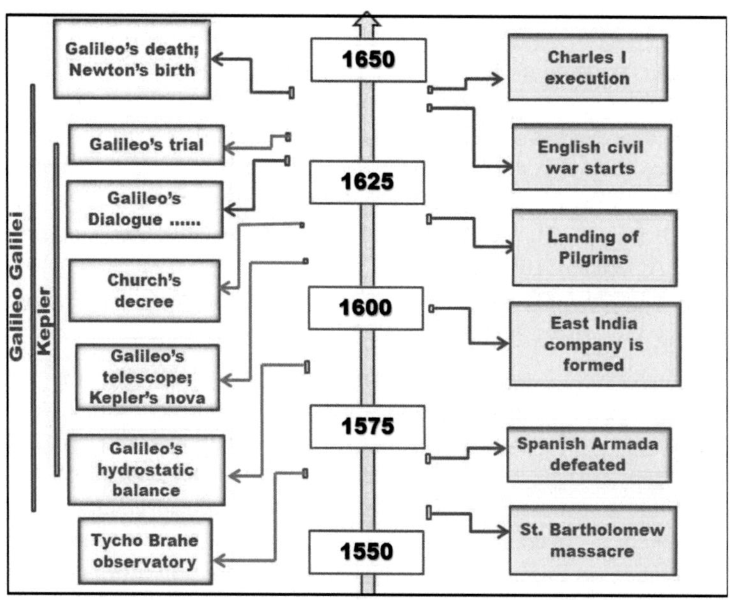

WHEN

Notes, References, and Credits

Notes and References

1. For more on Galileo, see, e.g.,
   Drake, Stillman (1990), *Galileo: Pioneer Scientist*, The University of Toronto Press, Toronto [ISBN 0- 8020-2725-3].
   Consolmagno, Guy and Schaefer, Marta (1994), *Worlds Apart, A Textbook in Planetary Science*, Englewood, Prentice-Hall, New Jersey [ISBN 0-13-964131-9].
   Chalmers, Alan Francis (1999), *What is this thing called Science?*, University of Chicago Press, Chicago [ISBN 978-0-7022-3093-6].

Blackwell, Richard J. (2006), *Behind the Scenes at Galileo's Trial*, University of Notre Dame Press, Notre Dame [ISBN 0-268-02201-1].
Gingerich, Owen (1992), *The Great Copernican Chase and other adventures in astronomical history*, Cambridge University Press, Cambridge [ISBN 0-521-32688-5].
Seeger, Raymond J (1966), *Galileo Galilei, his life and his works*, Pergamon Press, Oxford [ISBN 0-08- 012025-3].

2. See, e.g., https://www.physics.rutgers.edu/ croft/ANAGRAM.htm
3. For more on Galileo's encounter with the church, see:
   Feldhay, Rivka (1995), *Galileo and the Church: Political Inquisition Or Critical Dialogue?*, Cambridge University Press, Cambridge [ISBN 9780521344685].
   Finocchiaro, Maurice A (1989), *The Galileo Affair: A Documentary History*, University of California Press, Berkeley [ISBN 0-520-06662-6].
   Graney, Christopher M. and Danielson, Dennis (2014), *The Case Against Copernicus*, Scientific American **310** (1), 72–77.
4. See, e.g., Anton Postl (1997), *Correspondence between Kepler and Galileo, Vistas in Astronomy*, Vol. **21**, pp. 325–330.
5. It is not really clear how this phrase originated. The words "E pur si muove" were found in 1911 on a Spanish painting which was completed within a year or two of Galileo's death, as it is dated 1643 (or 1645, the last digit was smudged). The painting is not factually correct, because it depicts Galileo in a dungeon, but shows that some variant of the "Eppur si muove" anecdote was in circulation even immediately after his death. It had been circulating for over a century before it was published. The event was first mentioned (in English) in 1757 by Giuseppe Baretti in his book. The book was written 124 years after the supposed quote and became widely known:
   Baretti, G. M. A. (1757), *The Italian Library. Containing An Account of the Lives and Works of the Most Valuable Authors of Italy. With a Preface, Exhibiting The Changes of the Tuscan Language, from the barbarous Ages to the present Time*, A. Millar, in the Strand, London.

Figure Credits

6. Figure 14.1 courtesy: Justus Sustermans / Wikimedia Commons, Public domain. https://commons.wikimedia.org/wiki/File:Justus_Sustermans_-_Portrait_of_Galileo_Galilei,_1636.jpg (from public domain).
7. Figure 14.2 courtesy: https://commons.wikimedia.org/wiki/File:Galileo_telescope_replica_(1).jpg (from public domain).
8. Figure 14.3 courtesy: Giuseppe Bertini [Public domain], via Wikimedia Commons. https://commons.wikimedia.org/wiki/File:Bertini_fresco_of_Galileo_Galilei_and_Doge _of_Venice.jpg (from public domain).

15

Affairs of the Heart

> "The blood current flows continuously in a circle and never stops. The heart regulates all the blood of the body. [The flow of blood] may be compared to a circle without beginning or end."
>
> — Nei jing, ancient Chinese medical treatise (about 2500 BC)

The ancient Chinese text above shows that, since antiquity, men of medicine have known about the important role played by the heart in the human body and that it was 'somehow' connected with the flow of blood. But the understanding of precisely what this connection was proved to be a stumbling block in the advance of physiology for a long time [1], and it has an interesting history.

The earliest known writing about the circulatory system is in the Ebers Papyrus (about sixteenth century BC), which is an ancient Egyptian papyrus dealing with medicine, containing over 700 prescriptions and remedies (including some spiritual ones!). This papyrus recognizes the connection of the heart to the arteries. Egyptians thought that air came in through the mouth and went into both the lungs and heart. From the heart, the air is supposed to have travelled all over the body through the arteries. While this is only partially correct, it represents the earliest known account of scientific thought on this matter.

Around the sixth century BC, the circulation of vital fluids through the body was known to Sushruta, a physician in India. He also seems to have known about the arteries, described as 'channels'. The valves in the heart were known to the

T. Padmanabhan and V. Padmanabhan, *The Dawn of Science*,
https://doi.org/10.1007/978-3-030-17509-2_15

physicians of the Hippocratic school around the fourth century BC, but their function was not understood by them. Most of the early Greek scientists, with their blind devotion to Aristotle, believed that blood vessels contained both blood and air. They thought that the veins carried blood and the arteries carried mostly air. In fact, the word ‘artery’ itself originates from the Greek words meaning ‘air duct’.

It was Galen, the second century Greek physician, who conclusively demonstrated that the arteries contained only blood and identified venous (dark red) and arterial (brighter and thinner) blood, each with distinct and separate functions. But he believed that the air entered from the lungs into the right side of the heart and that the flow of blood in the vessels was like tides in the sea — ebbs and flows moving back and forth. The force that prodded this flow came from the contractions of the arteries, without the heart playing any significant role. Blood, it was thought, was produced in the liver from where it passed to the right auricle (one of the two upper chambers of the heart) and then to the right ventricle (one of the two lower chambers). Then the blood was supposed somehow to have made its way to the left side, where it met blood from the arteries which contained air coming from the lungs. These notions about blood circulation and the heart held sway for an incredibly long time — nearly fourteen centuries.

The first indication that this picture was inadequate came in the sixteenth century. The anatomical works of the Italian physician, Vesalius, and others like Avicenna, showed that there were no holes in the partition of the heart. So the flow of blood from the right side of the heart to the left became even more of a mystery. The solution to this mystery, however, *already existed in a book [2] by the Arabic scholar, Ibn al-Nafis (1213–1288), published as early as AD 1242!* Ibn al-Nafis — whom we mentioned earlier, in Chap. 6 — was the first person to accurately describe the process of pulmonary circulation, and for this reason he is sometimes considered to be the father of circulatory physiology.

In his work, *Commentary on Anatomy in Avicenna’s Canon*, Ibn al-Nafis stated that “the blood from the right chamber of the heart must arrive at the left chamber, but there is no direct pathway between them. The thick septum of the heart is not perforated and does not have visible pores as some people thought, or invisible pores as Galen thought. The blood from the right chamber must flow through the vena arteriosa (pulmonary artery) to the lungs, spread through its substances, be mingled there with air, pass through the arteria venosa (pulmonary vein) to reach the left chamber of the heart, and there form the vital spirit.” (This circular flow of blood from the heart to the lungs and back is now known as the ‘lesser circulation’.) In addition, Ibn al-Nafis also had an insight into what would later be identified as the capillary circulation. He said, “there must be small communications or pores (called *manafidh* in Arabic) between the pulmonary artery and vein," a prediction that anticipated the discovery of the capillary system by more than 400 years (see Chap. 18).

But it appears that the work of Ibn al-Nafis was not known to western scholars of the medieval period. So the 'lesser circulation' was to be independently rediscovered by a Spanish physician, Miguel Serveto (1511–1553) (also called Michael Servetus). But he published it in a theological treatise, *Christianismi Restitutio*, and not in a book on medicine. Unfortunately, the book also contained strong unitarian theological views and this got Servetus into trouble with John Calvin (1509–1564), who preached a much more extreme version of Protestantism than even Martin Luther (1483–1546) did [3]. After the publication of his book, Servetus had travelled to Geneva, which at that time was under Calvin's rule. Calvin immediately had Servetus arrested and burnt at the stake with all the copies of the book — which again tells us that religious reformists may not always help the progress of science.

Box 15.1: Blood Circulation: A Burning Issue

In 1553, Michael Servetus (also known as Miguel Serveto) published a religious work with strong anti-Trinitarian views titled, *Christianismi Restitutio* (The Restoration of Christianity). This work sharply rejected the idea of predestination in certain forms. *Incredibly enough, it also included a description of the pulmonary circulation!*

John Calvin, who had written his summary of the Christian doctrine, *Institutio Christianae Religionis* (Institutes of the Christian Religion), found Servetus' book to be an attack on the historical Nicene Christian doctrine and a misinterpretation of the biblical canon. Acrimonious debate ensued between Calvin and Servetus in the form of an exchange of letters. On 16 February 1553, Michael Servetus was denounced as a heretic by Guillaume de Trie, a rich merchant who was a good friend of Calvin's and who had taken refuge in Geneva. On behalf of the French inquisitor Matthieu Ory, Michael Servetus and Balthasar Arnollet, who had published *Christianismi Restitutio*, were questioned, but they denied all charges and were released due to lack of evidence.

Ory wanted definitive proof and, on 26 March 1553, he obtained the letters sent by Servetus to Calvin and some pages of *Christianismi Restitutio*. On this basis, Servetus was arrested on 4 April 1553 by the Roman Catholic authorities, and imprisoned in Vienne, France. Interestingly enough, he escaped from the prison three days later. On 17 June, he was convicted of heresy, in absentia, thanks to the 17 letters provided by John Calvin, and was sentenced to be burned at the stake with his books.

Intending to flee to Italy, Servetus inexplicably stopped in Geneva! He was arrested on 13 August and was again imprisoned. At his trial, Servetus was condemned on two counts, for spreading and preaching non-trinitarianism and anti-paedobaptism (anti-infant baptism). After several twists and turns, Servetus was sentenced, on 24 October, to death by burning. On 27 October, Servetus was burnt alive — atop a pyre of the copies of his own book which had the description of the blood circulation — at the Plateau of Champel near the border of Geneva. Only three copies of the book survived, but these were kept hidden for decades, for obvious reasons.

Fortunately, the concept (of lesser circulation) soon surfaced again and spread across Europe when an Italian anatomist, Realdo Colombo (1516–1559), independently re-rediscovered it and gave it prominence in his lectures [4]. Nevertheless, these 'new' ideas, combined with the fact that there were no perforations in the walls of the heart, were proving to be difficult to reconcile with Galen's views, which still dominated the scene. And the dogmatists persisted.

The next disturbing piece of evidence against Galen's ideas came from research done by Hieronymus Fabricius (1537–1619), who was a professor of medicine at the University of Padua in Italy. Fabricius studied the structure of the veins in detail and found that they contained a series of valves, the function of which was totally unclear. (Incidentally, Fabricius was a student of Fallopius (1523–1562), who discovered the Fallopian tubes; Fallopius himself was a student of Vesalius, thereby maintaining the strong tradition of medical science teaching at Padua.) Fabricius was so close to discovering the circulation of blood but, being a strong Galenist, he did not push his observations to their logical conclusion.

That honour [5] eventually went to the English physician, William Harvey (1578–1657). Harvey, being the son of a very prosperous businessman, had the best of educations. He received his first degree in medicine from the University of Cambridge in 1597. Determined to pursue the study of medicine further, he went to the best school at that time, viz., the University of Padua. He spent a couple of years in Padua working with Fabricius. Returning to London, Harvey rose very fast, both on the social and the professional ladder. He became a fellow of the College of Physicians, a practising physician at St. Bartholomew Hospital, and personal

Fig. 15.1: Portrait of William Harvey (1578–1657), who came up with the clearest exposition of blood circulation, as we understand it today [6]. The earliest specific reference [1] to heart and blood circulation is probably in the Chinese text (about 2500 BC) quoted at the beginning of this chapter. Later on, the concept of 'lesser circulation', viz., the circular flow of blood from the heart to the lungs and back, was described explicitly in a book by the Arabic scholar, Ibn al-Nafis, published as early as 1242 AD, but this had gone unnoticed. It was rediscovered by a Spanish physician, Michael Servetus (1511–1553), whose book on anatomy and blood circulation, unfortunately, also contained unitarian theological views. This got him into trouble with John Calvin, who had him burnt at the stake with almost all extant copies of his book. Years later, the Italian anatomist, Realdo Colombo (1516–1559) re-rediscovered it and another Italian, Fabricius (1537–1619), came very close to discovering the full blood circulation. Harvey deserves credit for coming up with the clearest description of blood circulation.

physician to King James I and later to King Charles I, in quick succession. Being very courteous and dignified, he was held in affection and respect by his colleagues and was successful in every way.

In spite of active medical practice, Harvey found time to pursue research over a long period (1604–1642). The two years with Fabricius early on in his career had convinced Harvey that the flow of blood in the body was not well understood, and he started a series of simple experiments and dissections to get to the bottom of the problem. By dissection, he realized that there were valves separating the auricles from the ventricles, so that blood could flow only from the auricle to the ventricle. He had earlier learnt from Fabricius about the valves within the veins. Combining these observations, he could arrive at the complete picture of blood circulation.

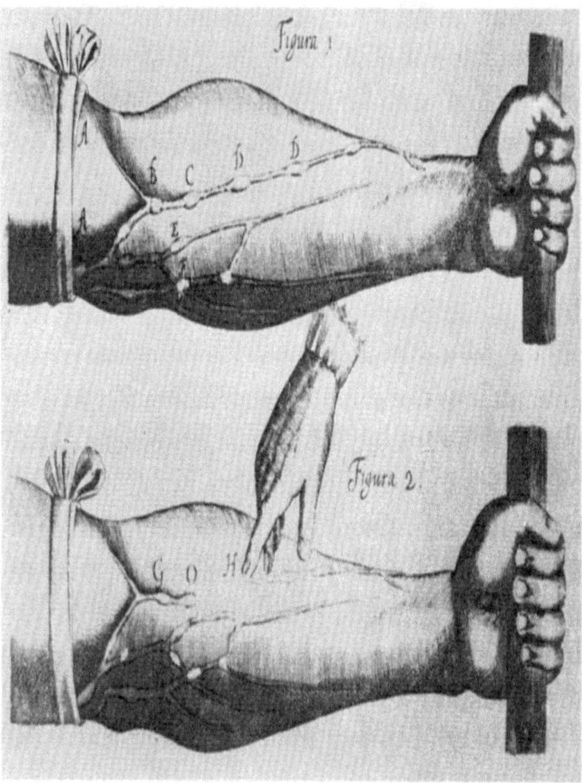

Fig. 15.2: Harvey's demonstration that blood flowed only in one direction in the veins by pressing on the vein and releasing his fingers [7]. He showed that, when he released the upper finger, the blood would not flow down.

Claiming that a fixed volume of blood circulated through the blood vessels and the heart in a definite direction was quite a revolutionary idea. Harvey also found simple and elegant arguments to demonstrate the fact that blood flowed only in a definite direction within the body. For example, he showed that, when an artery was tied off, it bulged on the side towards the heart; whereas if a vein was treated in the same way, it bulged on the side away from the heart (see Fig. 15.2). Harvey also noted, from a straightforward estimate, that the quantity of blood pumped out by the heart in one hour was roughly equal to three times the weight of a man. Once again, this strongly suggested the *circulation* of a fixed amount of blood.

Harvey published these results in 1628 in the book *Exercitatio Anatomica de Motu Cordis et Sanguinis in Animalibus* (An anatomical exercise on the motion of the heart and the blood in living beings), commonly refered as *De Motu Cordis*. It was a small book, containing only about 72 pages, printed in Holland on very poor paper and full of typographical errors. All the same, the book was destined to become a scientific classic. The Galenists of the day, of course, ridiculed Harvey, but his forceful arguments eventually convinced his colleagues of his new ideas.

Box 15.2: The Circulator

Scientists who come up with ideas that go counter to popular beliefs risk the fate of being ridiculed (sometimes meted out with more serious forms of opposition). Harvey was no exception, as Galen's followers did not readily accept his ideas.

An immediate consequence was that business at his clinical practice went down sharply! Common patients were justifiably hesitant to go to a doctor whose strange views were jeered at by others in the profession! In fact, some of his critics even called him the 'circulator', not because he was campaigning for the circulation of blood, but because it was a Latin slang for 'quack', usually attributed to peddlers who were hawking medicines for all sorts of diseases at the circus! Though it did take time to re-establish his reputation, Harvey managed it by quietly and calmly defending his work.

The theory of the circulation of blood depended crucially on the answer to one question: how did blood flow from the arteries to the veins? Noting that the arteries and veins branched into finer and finer vessels in the body, Harvey had the intuition to guess that the transfer occurred at the finest levels (capillaries) which are too small to see. This was confirmed later by the Italian physiologist, Marcello Malpighi (1628–1694), using a microscope (see Chap. 18).

Harvey was a good friend of King Charles, but had the sense to stay clear of politics. Thanks to this, he escaped the wrath of the parliamentary army when the Civil War started in England in 1642 and Charles I was arrested. The beheading of Charles in 1649, however, affected Harvey quite deeply. This loss, and the fact that Cromwell (the leader of the parliamentary forces) always treated Harvey as a possible suspect, made Harvey quite an unhappy man in his later years. He died in 1657 at the age of 80.

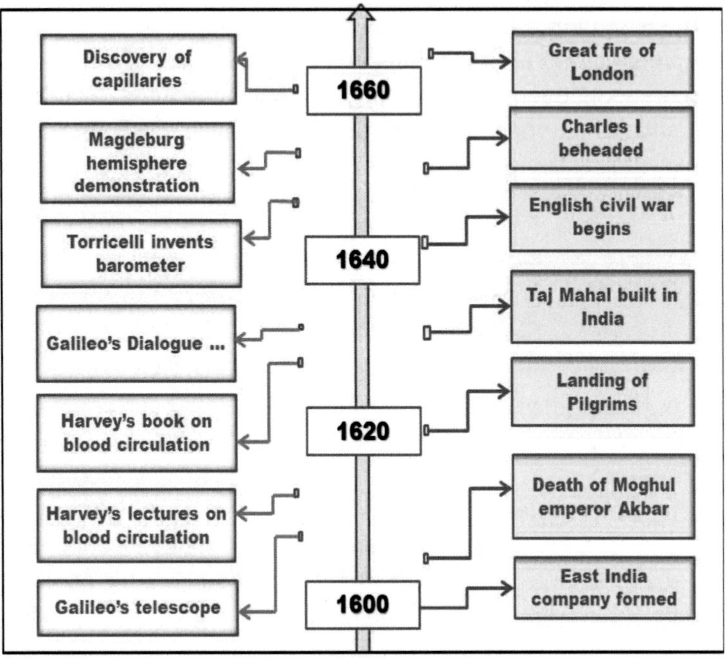

WHEN

Notes, References, and Credits

Notes and References

1. See,
   Zhu Ming (tr) (2001), *The Medical Classic of the Yellow Emperor*, Foreign Language Press, Beijing, China [ISBN 7-119-02664-X].
   There is, however, some dispute about the Chinese text. For an alternative point of view, see, e.g.,
   Nie Jing-Bao (2001), *Refutation of the claim that the ancient Chinese described the circulation of blood: a critique of scientism in the historiography of Chinese medicine*, New Zealand Journal of Asian Studies **3** (2), 119–135.
2. See, e.g.,
   Loukas M., et al. (2008), *Ibn al-Nafis (1210–1288): The first description of the pulmonary circulation*, Am. Surg. **74** (5), 440–442.
   Haddad, Sami and Amin A. Khairallah (1936), *A Forgotten Chapter in the History of the Circulation of Blood*, Annals of Surgery **104**(1), 1.
   Said, Hakim Mohammed (1994), *Knowledge of the circulation of the blood from Antiquity down to Ibn al-Nafis*, Hamdard Medicus **37**, (1), 5–37.
3. See, e.g.,
   Bainton, Roland H. (2005), *Hunted Heretic: The Life and Death of Michael Servetus 1511–1553*, Blackstone Editions, USA [ISBN 0-9725017-3-8].
   Lovci, Radovan (2008), *Michael Servetus, Heretic or Saint?*, Prague House, Prague [ISBN 1-4382-5959-X].
   Nigg, Walter (1990), *The Heretics: Heresy Through the Ages*, Dorset Press, UK [ISBN 0-88029-455-8].
   Goldstone, Nancy Bazelon and Goldstone, Lawrence (2003), *Out of the Flames: The Remarkable Story of a Fearless Scholar, a Fatal Heresy, and One of the Rarest Books in the World*, Broadway, New York [ISBN 0-7679-0837-6].
4. See, e.g., Cunningham, Andrew (1997), *The Anatomical Renaissance: The Resurrection of the Anatomical Projects of the Ancients*, Scholar Press, The British Journal for the History of Science **31**(4), 469–487 [ISBN 1-85928-338-1].
5. For more on Harvey and his contributions, see:
   Gregory, Andrew (2001), *Harvey's Heart, The Discovery of Blood Circulation*, Icon Books, Cambridge, England.
   Wright, Thomas (2012), *Circulation: William Harvey's Revolutionary Idea*, Chatto, London.

Figure Credits

6. Figure 15.1 courtesy: Daniel Mijtens [Public domain], via Wikimedia Commons. https://commons.wikimedia.org/wiki/File:William_Harvey_2.jpg (from public domain).
7. Figure 15.2 courtesy: Unknown/ [Public domain] via Wikimedia Commons. https://commons.wikimedia.org/wiki/File:William_Harvey_%281578-1657%29_Venenbild.jpg (from public domain).

16

The Weight of the Intangible

Though air is a pervasive and powerful agent existing all around us, a serious understanding of how air (and gases in general) behave came about only in the seventeenth century. This involved the collective efforts of several people, and especially of Jan Baptista van Helmont (1580–1644) from Belgium, Otto von Guericke (1602–1686) from Germany, Evangelista Torricelli (1608–1647) from Italy, and Blaise Pascal (1623–1662) from France.

Of these, Helmont's work is not as well known as it should be [1]. This is a bit unfair, because it was he who first recognized the important fact that air was not a unique entity (as the Greeks believed) and that many 'air-like' substances could be produced by ordinary chemical processes.

Like several others before him (notably Paracelsus), Helmont was a physician who was also interested in chemistry. Most of his work — again like that of the others — revolved around finding a 'philosopher's stone' and other alchemical endeavours, which meant that it was fairly worthless. However, in the midst of this nonsense, he did actually discover something important.

He noticed that his experiments produced several 'vapours', which had no definite shape (and took the shape of the containers in which they were kept), behaving exactly like air physically, but having very different chemical properties compared to normal air. In particular, he made a detailed study of the vapour he could produce by burning wood. He called it 'gas sylvestre', which was Flemish for 'gas from wood'. (Incidentally, he also coined the term 'gas', which essentially meant 'chaos' in Flemish, but the term did not catch on for another century and was

T. Padmanabhan and V. Padmanabhan, *The Dawn of Science*,
https://doi.org/10.1007/978-3-030-17509-2_16

reintroduced by Lavoisier (1743–1794)!) Helmont listed the properties of this gas and stressed the fact that it was indeed quite different from air. One might say that this was the first time that the three states of matter (solid, liquid, and gas) were clearly recognized.

Fig. 16.1: Otto von Guericke (1602–1686), who was the first to understand the effects of the vacuum with any clarity, led an interesting life [5]. He started out as a professional engineer, entered politics in 1627, right in the middle of the Thirty Years' War, and barely escaped with his life during the sack of Magdeburg in 1631. He then joined the army of Gustav II Adolphus of Sweden and returned to Magdeburg in 1632 to rebuild it, and became its mayor in 1646! Most of his scientific contributions, related to gas pressure, came later on.

The life and work of Guericke, by comparison, were a lot more colourful [2] than Helmont's. An engineer by profession, Guericke entered politics in 1627 right in the midst of the Thirty Years' War. His home town, Magdeburg (in Germany), was sacked by the imperial army in 1631, destroying most of the town, and Guericke barely managed to escape. He joined the army of Gustav II Adolphus of Sweden who actually turned the tide of the war. Guericke returned to Magdeburg the following year and, with his engineering background, led the process of rebuilding the town, becoming its mayor in 1646.

Meanwhile, Guericke got interested in the possibility of producing and maintaining a vacuum. The prevailing view on the subject was once again Aristotle's: *Nature abhors a vacuum*. Aristotle had worked out a theory of motion in which a given body would move faster if the surrounding medium became less dense. In his theory, bodies had to move with infinite speed in vacuum, which Aristotle thought was impossible; hence, he concluded that a vacuum could not exist. Guericke was not impressed by this argument and decided to settle it by direct experiment. So he built an air pump, similar in design to the water pump which had been used for centuries, and used it to evacuate air from a closed chamber.

Box 16.1: Vacuous Showmanship?

Guericke had a natural flair for showmanship and used his air pump for some dramatic demonstrations. He first tied a rope to a piston and had 50 strong men pull the rope while he produced a vacuum on the other side with his pump. Very soon, the 50 men could not pull the piston up, working against what Guericke called *the force of the vacuum* (see Fig. 16.2).

He then went on to perform an even more dramatic experiment, involving the famous *Magdeburg hemispheres*. These were two hemispheres, which were fitted together perfectly along a greased, air-tight edge. When the hemispheres were put together and the air inside was evacuated, the hemispheres could not be separated even by two teams of horses trying to pull them apart (see Fig. 16.3). And as soon as he let air back into the spheres, they fell apart by themselves. Guericke, the showman that he was, arranged a demonstration for the Emperor, Ferdinand III, in 1654. Needless to say, the Emperor was duly impressed.

Did Guericke understand the physics behind his experiments? There is no clear evidence that he did! That explanation came from Torricelli, a mathematician from the University of Rome [3].

He showed that candles would not burn and animals could not live in such an evacuated vessel. After these simple demonstrations, he decided to do something more dramatic. His experiments involving strong men and horses (Figs. 16.2 and 16.3) working against the force of the vacuum (as he called it) might not have

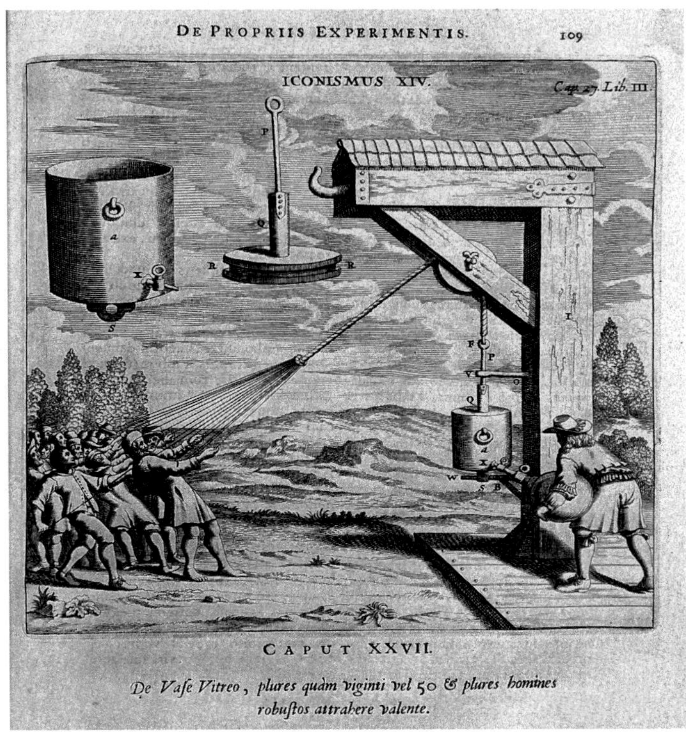

Fig. 16.2: Dramatic display of the power of the vacuum by Otto von Guericke, which caught the imagination of the public and the royalty [6]. After showing that candles would not burn and animals could not live inside an evacuated vessel, he decided to demonstrate the power of the vacuum more vividly. He tied a rope to a piston and arranged for 50 strong men to pull the rope as he produced a vacuum on the other side with his air pump. The men could not pull the piston up against what Guericke called ‘the force of the vacuum’.

had major scientific content, but they caught the attention of the royalty and the imagination of the public!

Did Guericke understand the reason why these demonstrations were successful? Maybe he did, but if so, he never described it; a clear enunciation of the basic principles came from Torricelli, a mathematician from the University of Rome [3]. Torricelli was deeply attracted by Galileo’s work. So much so that he went to meet Galileo and volunteered to serve as his secretary during the last three months of

Fig. 16.3: Yet another dramatic demonstration of 'the force of the vacuum', involving the famous *Magdeburg hemispheres* [7]. These two hemispheres were fitted together perfectly along a greased edge. When the air inside was evacuated, the hemispheres could not be separated even by two teams of horses trying to pull them apart. But when the air was allowed back into the spheres, they fell apart by themselves. Guericke arranged a demonstration for the Emperor, Ferdinand III, in 1654 and, needless to say, the Emperor was very impressed.

Galileo's life. During these three months, Galileo pointed out to him a peculiar problem which needed explanation.

The problem was the following. Pumps used for raising water from one level to another were made using pistons. Their action was explained on the basis of Aristotle's principle: nature abhors a vacuum, so when the piston is raised, a vacuum would be created inside the pump unless the water level inside rises; and to avoid the vacuum, the water rushes in. There was, however, one problem with this explanation.

People who used these pumps knew that they could raise the water only up to a height of about 33 feet (10 metres). Even when longer pumps were used, the water would rise up to just about 33 feet and then stop (see Fig. 16.4). Galileo found it strange that nature abhorred a vacuum up to a certain limit, but gave up tamely after that. He suggested that Torricelli look into this strange behaviour.

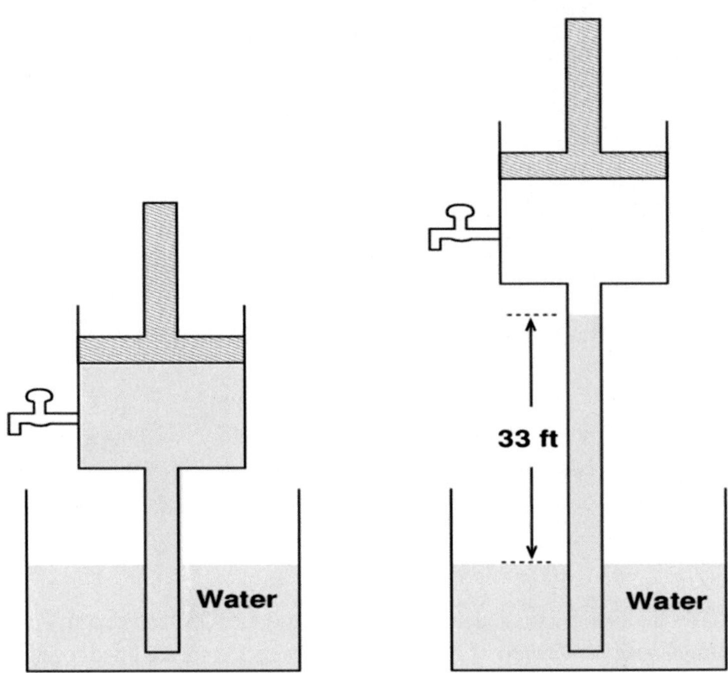

Fig. 16.4: A clue to the nature of air pressure. Left: Water rises inside the tube when the piston is moved up. Right: However, it will not rise above 33 feet, even when the piston is raised higher. This shows that the atmospheric air column could only support the pressure of 33 feet of water.

It occurred to Torricelli that the entire phenomenon had a much simpler mechanical explanation: "Suppose the air above us had some weight," reasoned Torricelli, "then it will push the water surface and make it rise inside the pump when the piston is raised. However, suppose the total weight of the air could only support the weight of about 33 feet of water. Then no pump with a piston can raise the water higher. All this talk about 'vacuum abhorrence' is irrelevant."

Torricelli actually went further. Since mercury was about 13.5 times heavier than water, air could only support a column of mercury which was 13.5 times smaller compared to water. This works out to (33 feet/13.5) or about 30 inches (76 cm). Torricelli took a four-foot long tube which was closed at one end, filled it with mercury, put his thumb to close the other end and inverted the tube into a large dish of mercury. On releasing the thumb, mercury flowed from the tube into the dish, but not all of it. Nearly 30 inches of mercury remained in the tube, proving Torricelli's ideas to be correct.

This was in 1643 (seven years before Guericke's invention of the air pump) and must be considered a milestone in science. Not so much for the intrinsic importance of the result, but for one of the clearest and simplest demonstrations of what we now call the 'method of science'. The key first step was in realizing that a specific, unexplained fact existed (viz., water pumps did not work for raising more than 33 feet of water). The second step was in postulating an explanation for the observed fact (that the air had a weight equivalent to that of 33 feet of water). Several other possible explanations could have been offered, but what distinguished the scientific one from the others was the third crucial step: *a prediction was made* using a hypothesis (nearly 30 inches of mercury could be supported by the weight of air). And finally, the prediction was directly tested and verified. The Aristotelian explanation, 'nature abhors a vacuum' was not only wrong but also deficient in predictive power. The part above the mercury in the inverted tube is a vacuum except for small amounts of mercury vapour. This was the first human-made vacuum, and is called the Torricelli vacuum in his honour.

These ideas suggest one more result: if it was the weight of the air that was supported by the mercury in the tube, then the mercury level should drop when a barometer (which was what Torricelli's tubes were called) was taken up a mountain. The verification of this claim was designed [4] by Blaise Pascal (1623–1662). Being a weak and chronically sick man, Pascal did not attempt the task himself but managed to persuade his brother-in-law to carry the barometers up the sides of the Puy de Dome in France in 1648. The helpful relative climbed a height of about a mile and found that the mercury had dropped by a few inches, according with the predictions.

As might be expected, Pascal also introduced, literally, the true French spirit into these studies. He made a barometer using red wine (which is lighter than water) and checked that a column of about 14 metres (46 feet) could be supported!

Box 16.2: The Rise of the Atom

The study of the behaviour of gases in different circumstances had an important consequence. It gradually made scientists turn their attention to the possible atomic nature of matter, something which we now consider to be fundamental.

The atomic view of matter, like several other fundamental concepts, has a chequered history. It appears that these ideas were originally suggested in the West by the Ionian philosopher Luecippus (around 450 BC) and his student Democritus. (Similar ideas with different levels of detail originated in other ancient civilizations as well, as in India, for instance.) Democritus was of the view that atoms of different elements have distinct shapes and sizes, and this distinction led to the different properties of these elements. Though this view was not accepted by most philosophers, atomism continued to exist, notably for the Greek philosopher Epicurus (341–270 BC) and others who embraced his philosophy.

This Epicureanism had many followers over the next few centuries, and one among them was the Roman poet Lucretius (99–55 BC). He probably wrote the first didactic poem — i.e., a poem whose purpose is to teach a concept — entitled *De rerum natura* (On the Nature of Things) around the first century BC. This long poem, explaining the Epicurean atomic nature of matter, was divided into six untitled books! Centuries later, the poem appeared in print in 1473, thanks to the advent of the printing press, and it became well known among European scholars.

There was, however, one rather ancient objection to this idea. It was argued that, if matter was made of atoms, then one could compress matter by moving these atoms closer together. But compressing a solid or liquid by any significant amount was far from easy, implying that the atoms — even if they existed — were already as closely packed as possible. Such a configuration would then be indistinguishable from a continuum and hence it seemed pointless to introduce the concept of atoms at all.

The study of the behaviour of gases dramatically changed this point of view. Following the work of Guericke and Torricelli, the Irish chemist Robert Boyle (1627–1691) investigated the behaviour of gases subjected to different amounts of pressure, in a fairly quantitative manner. This led to what is called Boyle's law, viz., that if you double the pressure on a gas you can compress it to half its volume. That is, the product PV remains a constant, where P is the pressure and V is the volume. This, of course, only holds true if the temperature of the gas is held constant, a fact that was clearly stated by the French physicist Edme Mariotte (1620–1684), who also independently discovered this law. As a result, continental Europe called it Mariotte's law for quite sometime.

The easy compressibility of gases clearly suggested that gases could be made of atoms with a fair amount of empty space in between. This could explain how, under pressure, the volume of a gas shrinks, because the atoms are brought closer together. This point of view was emphasised by the French philosopher Pierre Gassendi (1592–1655) and very soon Robert Boyle became a true believer in the atomic nature of matter. Thus, the investigations of gases led to far more significant concepts than one might have anticipated at first.

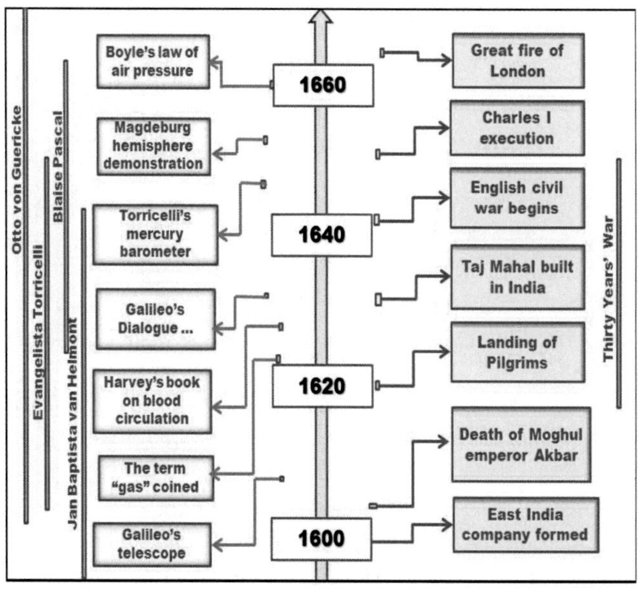

WHEN

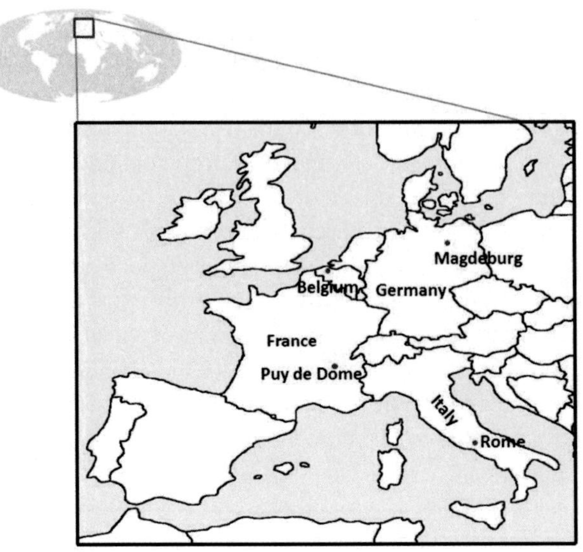

WHERE

Notes, References, and Credits

Notes and References

1. See, e.g.,
   Redgrove, I. M. L. and Redgrove, H. Stanley (2010), *Joannes Baptista van Helmont: Alchemist, Physician and Philosopher*, Kessinger Publishing, USA [ISBN 978-1169686885].
   Pagel, Walter (2002), *Joan Baptista van Helmont: Reformer of Science and Medicine*, Cambridge University Press, Cambridge [ISBN 978-0521526555].
2. See, e.g.,
   Conlon, Thomas E (2011), *Thinking About Nothing: Otto von Guericke and the Magdeburg Experiments on the Vacuum*, The Saint Austin Press, San Francisco [ISBN 978-14478-3916-3].
   Kleint, Christian (1998), *Horror, Happenings and Highlights in the History of Vacuum Physics*, Progress in Surface Science **59**, 301–312.
3. See, e.g.,
   Jervis-Smith and Frederick John (1908), *Evangelista Torricelli*, Oxford University Press, Oxford [ISBN 9781286262184].

Robinson, Philip (1994), *Evangelista Torricelli*, The Mathematical Gazette **78** (481), 37.

4. For more on Pascal, see, e.g.,
Connor, James A. (2006), *Pascal's Wager: The Man Who Played Dice with God*, HarperOne, USA [ISBN 978-0060766917].
McPherson, Joyce (1997), *A Piece of the Mountain: The Story of Blaise Pascal*, Greenleaf Press, Australia [ISBN 978-1882514175].

Figure Credits

5. Figure 16.1 courtesy: Anselm van Hulle [Public domain, CC0 1.0], via Wikimedia Commons.
https://commons.wikimedia.org/wiki/File:Anselmus-van-Hulle-Hommes-illustres_MG_0539.tif (published under CC0 1.0).
6. Figure 16.2 courtesy: Photo number: L0006004, Credit: Wellcome Collection.
https://commons.wikimedia.org/wiki/File:O._von_Guericke,_experiment_with_vacuum, _17th_century_Wellcome_L0006004.jpg (published under CC By 4.0).
7. Figure 16.3 courtesy: Gaspar Schott , [Public domain] , via Wikimedia Commons.
https://commons.wikimedia.org/wiki/File:Magdeburg.jpg [from public domain].

17

Geometry without Figures

There is a story about Ptolemy Soter (367–282 BC), the first king of Egypt and founder of the Alexandrian museum, who studied geometry under Euclid ($\sim$ fourth century BC). Ptolemy found the subject rather difficult and apparently asked his teacher whether there was an easier way of learning geometry. To which he received the famous reply: "There is no royal road to geometry" [1].

In a sense, Euclid was wrong. There is a way of doing geometry using algebra which is considerably more powerful and conceptually straightforward. The discovery of this connection between algebra and geometry, nearly nineteen centuries after Euclid, was definitely a milestone in science.

The person who contributed most to this subject was Rene Descartes (1596–1650), the French philosopher and mathematician [2]. Descartes used to sign in the Latinized version of his name, 'Cartesins', and because of this, his systems of geometry and philosophy both go under the name 'Cartesian'.

Descartes was born in France in 1596. He was a brilliant but very unhealthy student and obtained the concession from his teachers to remain in bed, as long as he wished, everyday — a habit which he continued into his adult years. His early Jesuit education made him extremely devout and faithful. In 1633, when he heard about Galileo's fate, he abandoned the idea of writing a book in support of the Copernican theory. Instead, he came out with an astronomical model in which the Earth was in the centre of a 'cosmic vertex' with the vertex travelling around the Sun. Though many people accepted this compromise, it was worthless as an astronomical model.

T. Padmanabhan and V. Padmanabhan, *The Dawn of Science*,
https://doi.org/10.1007/978-3-030-17509-2_17

Fig. 17.1: Portrait of Descartes (1596–1650), who created the branch of mathematics now called 'analytic geometry', which essentially reduces geometry to algebra [5]. Given his philosophical inclinations, he published his results in the last of the three appendices accompanying his French treatise *Discourse on the Method of Rightly Conducting One's Reason and of Seeking Truth in the Sciences*. Most of the treatise is philosophical and quite irrelevant; but the third appendix makes up for the rest. He died of pneumonia, caught perhaps as a consequence of having to give lectures at five in the morning, three times a week, to Queen Christina of Sweden, during one of Sweden's worst winters. For interesting political and religious reasons, Descartes' body achieved the dubious distinction of being buried three times.

After his education, Descartes joined the French army. Fortunately, he was never exposed to actual warfare and hence had plenty of spare time to work out his ideas. It was during this time that Descartes seems to have stumbled upon an important discovery, which created the branch of mathematics called 'analytic geometry'. The basic idea of analytic geometry was extremely simple. It was well known (right from the days when maps were first used) that the position of a point on a two-dimensional surface could always be represented by two numbers (see Fig. 17.2). For example, the location of any city on the Earth could be specified

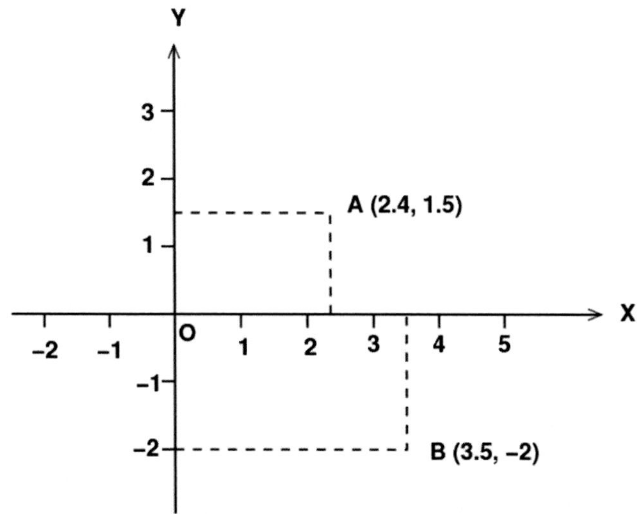

Fig. 17.2: Any point on a plane can be specified by giving two numbers. We first draw reference axes (OX, OY) and mark them at equal intervals. Distances to the 'left' or 'below' are treated as negative. The point A can be specified by its distance along the horizontal axis OX (2.4 units) and its distance along the vertical axis OY (1.5 units), in that order. Similarly, to denote a point below the OX axis (say B), we specify the corresponding distance along the vertical axis with a negative number; e.g, $(3.5, -2)$ in this case.

by the latitude and longitude. Similarly, the position of any point on this paper could be denoted by giving its distances from the bottom edge of the paper and from the left end. Descartes realized that this fact allowed any curve in geometry to be represented by an algebraic equation. For example, consider the equation $y = x^2/3$. If we now give for x a series of values like $1, 2, 3$, etc., we will obtain for y the values $1/3, 4/3, 3$, etc., respectively. Each of these pairs of numbers — that is, $(1, 1/3), (2, 4/3), (3, 3)$ — represents a point on the plane. By connecting these points, we obtain a smooth curve, which is unique (see Fig. 17.3). We have thus coded the information about the geometrical curve into the algebraic equation $y = x^2/3$. We can now methodically translate all the basic geometrical relations concerning the plane figures into equivalent algebraic statements. Once this is done, any property of a geometrical figure can be worked out using purely algebraic operations, without ever even drawing the figure.

The connection described above is extremely powerful and forms the cornerstone of applied geometry today. Its power stems from two facts. First, this connection

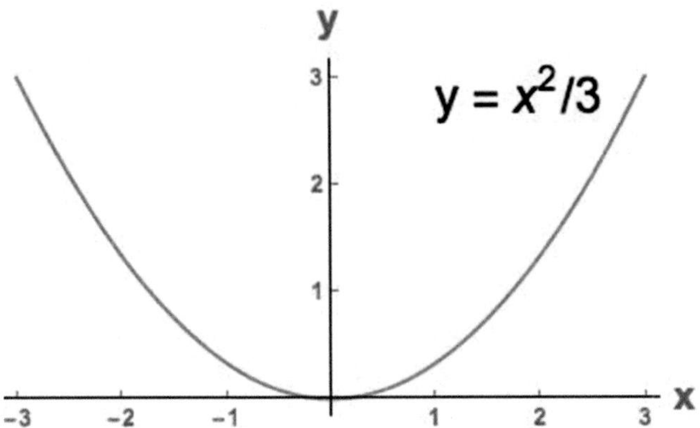

Fig. 17.3: Figures from equations, i.e., geometry from algebra. A correspondence can be established between a function, say, $y = x^2/3$, and a curve on the plane. The function provides a set of pairs of numbers; in the case of $y(x) = x^2/3$, $y = 0$ at $x = 0$; $y = 1/3$ at $x = 1$; $y = 4/3$ at $x = 2$; $y = 3$ at $x = 3$, etc. These sets of points can be connected by a smooth curve. In this manner, geometric curves can be expressed as algebraic functions and vice versa.

between algebra and geometry provides a systematic procedure for deriving and proving geometrical relations; for example, suppose we have to prove that the area of a triangle with sides a, b, c, is the square root of the expression $s(s-a)(s-b)(s-c)$, where s is half the perimeter of the triangle. In conventional geometry, there is no obvious way of proving such a result. One has to draw the figure of the triangle, introduce clever constructions and use appropriate theorems to arrive at this result. But analytic geometry provides a *brute-force* method for tackling the same problem. We can represent the three vertices of the triangle by three pairs of numbers — (x_1, y_1), (x_2, y_2), (x_3, y_3). By using the rules of analytic geometry — and without requiring much creativity or imagination — we can write down the expression for the area of a triangle and, comparing it with the given expression, we can easily prove the result.

The second reason why analytic geometry is so powerful is that it enormously extends the range of geometrical curves and shapes that can be studied. Conventional Greek geometry was restricted to straight lines, circles, conic sections, and some simple extensions of these. But now every equation of the form $y = f(x)$ could be represented by some curve in the plane and its properties could be studied. In this sense, algebra could enrich the scope of geometry.

Descartes was led to this discovery through his mechanistic view of the world. He tried to think of all the phenomena in terms of the motion of mechanical gadgets; to him, a geometrical curve was essentially a path traced by an object moving in a particular way. Descartes published his results as the last of the three appendices accompanying his treatise, *Discours de la méthode* ... — the full title of the book is a mouthful! A rough translation would be "Discourse on the Method of Rightly Conducting One's Reason and of Seeking Truth in the Sciences"! The book dealt with the proper way of conducting logical study in various branches of science. Most of the material in the treatise is trivial and useless; however, the third appendix contains the real gem. (It is rather amusing to note that there has been at least one other occasion in mathematics when the appendix was more important than the text. In 1832, a Hungarian mathematician, Farkas Bolyai, published a textbook on mathematics in which he included a 26 page appendix written by his twenty year old son, Janos Bolyai. This appendix contained the foundations of non-Euclidean geometry which was, of course, far more important than anything else in the book!)

After his service in the army, Descartes returned to Protestant Holland where he stayed almost till the end of his life. During this period, Queen Christina ruled Sweden and she was one of the most eccentric queens who ever lived. It had become fashionable in Europe at this time for the royalty to invite intellectuals to their courts and pretend a keen interest in matters of the mind. Following this trait, Christina invited Descartes to the Swedish court, which — unfortunately — he accepted. Christina made Descartes call on her three times a week at five in the morning, during one of the worst Swedish winters. It is not known whether the queen grew in intellect as a result of this exercise; poor Descartes, however, caught pneumonia and died.

Box 17.1: The Afterlife of Descartes

Descartes' afterlife was quite possibly as interesting as his life [3]. Queen Christina originally wanted to bury the great philosopher in Stockholm with full pomp and honors, no doubt hoping that he could add some lustre to her court in death that he hadn't been able to in his life. It was, however, pointed out to her that Descartes was a French Catholic; France, which was a Catholic nation, might not quite appreciate one of its native sons being buried in a Lutheran ambience. So poor Descartes was quietly buried in

Protestant Sweden, in a cemetery reserved for unbaptized children at Adolf Fredrikskyrkan in Stockholm.

Later on, in 1667, his remains were taken to France and buried again in the church of Sainte-Geneviève-du-Mont in Paris. Once again, the event was originally planned with a certain amount of pomp and splendour and a public funeral oration, but at the last minute an order came from Louis XIV's government forbidding it. So the tone had to be subdued for Descartes' second burial as well. (Incidentally, in 1663, the Pope placed Descartes' works on the Index of Prohibited Books because his views were considered 'too hot'!)

During the French Revolution, his remains came in for scrutiny for a third time! The church of Sainte-Geneviève was one among hundreds of religious sites seized by the state during the French Revolution, and the idea was floated that the remains of great men should occupy a Pantheon that was yet to be built. In the midst of the chaos and confusion of the Revolution — and the debate as to who these great men should be — the National Convention in 1793 finally published a decree to "accord to Rene Descartes the honors of the great Men, and order the transfer of his body to the French Pantheon"! The village in the Loire Valley, where he was born, was renamed La Haye-Descartes in 1802, then shortened to Descartes in 1967. His tomb is currently in the church of Saint-Germain-des-Prés in Paris, except for his cranium, which is in the Musée de l'Homme.

Almost at the same time, there was another Frenchman who independently developed parts of analytic geometry. This was Pierre de Fermat (1607–1665), a parliamentary counsellor who only did mathematics in his spare time [4]. He worked in mathematics 'for fun' and developed the frustrating habit of stating theorems on the margins of books and in private correspondence. Around 1630, he developed analytic geometry in two and three dimensions, but never bothered to publish it. We know of this only from a letter he wrote to a friend in 1637.

Box 17.2: Fermat's Last Theorem

Fermat (1607–1665) was a contemporary of Descartes and is often called the 'prince among amateur mathematicians'. He was a counsellor at the French Parliament, but devoted much of his spare time to mathematics. He made contributions in developing the theories of probability and analytic geometry and had some notions of calculus. But what probably made him 'immortal' was his contributions to the theory of numbers — in particular, a note he scribbled on the margin of a book, viz., *Arithmetica* of Diophantus.

This marginal note said that an equation of the form $x^n + y^n = z^n$ cannot be satisfied with x, y, z and n being integers with $n > 2$. (When $n = 2$, there are several solutions like $3^2+4^2 = 5^2$ and $5^2+12^2 = 13^2$, etc.) Fermat added that "I have discovered a truly marvellous proof of this, which, however, the margin is not large enough to contain." Many a later mathematician wished that *Arithmetica* had had a wider margin.

Fermat also stated several other 'theorems' in his correspondences, and in all but one case he was proved right (in one case he was proved wrong). Only the 'last theorem', quoted above, stood without being proved right or wrong for a long time. A tremendous amount of research went into investigating this theorem and new vistas of mathematics have originated from these studies. This theorem also led to dozens of false proofs year after year. Since the statement of the theorem is so simple, it has always been accessible to anyone with a basic education in algebra. Many amateurs tried their hand at it without even appreciating the true problem. To be fair to them, it must be said that even some professional mathematicians have been guilty of publishing wrong proofs!

In June 1993, the British mathematician Andrew Wiles presented what he considered to be the proof of *Fermat's Last Theorem* using fairly sophisticated mathematical techniques. Interestingly enough, a critical portion of the proof contained an error which was picked up by several mathematicians refereeing Wiles' manuscript. Wiles, along with his former student Richard Taylor, had to spend nearly a year to take care of this difficulty, and the final result was submitted to journals in the form of two papers (one by Wiles and the other co-authored with Taylor) in October 1994 and published in 1995, about 358 years after Fermat first conjectured it!

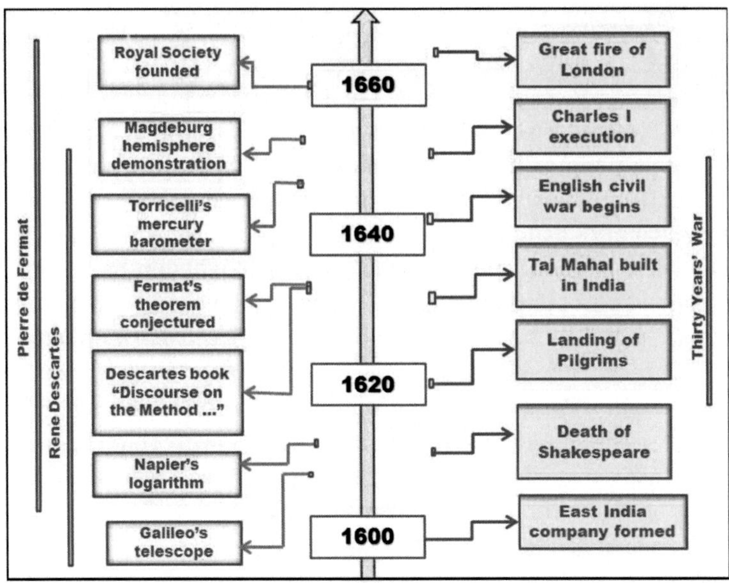

WHEN

Notes, References, and Credits

Notes and References

1. One source for this quote is Proclus (412–485 AD) in his Commentary on the First Book of Euclid's Elements. A translation of this text is available in:
   Taylor, Thomas (1788), *The Philosophical and Mathematical Commentaries of Proclus on the First Book of Euclid's Elements, and his Life by Marinus. With a preliminary Dissertation on the Platonic Doctrine of Ideas*, Vol 1, p. 101.
   Another, more accessible, translation which has this quote is:
   Morrow, Glenn R. (Translator) (1970), *Proclus: A Commentary on the First Book of Euclid's Elements*, Princeton University Press, Princeton, p. 57 [ISBN 978-0691071602].
2. For more on Descartes' life and times, see, e.g.,
   Clarke, Desmond (2006), *Descartes: A Biography*, Cambridge University Press, Cambridge [ISBN 0-521-82301-3].
   Cottingham, John (1992), *The Cambridge Companion to Descartes*, Cambridge University Press, Cambridge [ISBN 0-521-36696-8].
   Garber, Daniel (1992), *Descartes' Metaphysical Physics*, University of Chicago Press, Chicago [ISBN 0-226-28219-8].

Gaukroger, Stephen (1995), *Descartes: An Intellectual Biography*, Oxford University Press, Oxford [ISBN 0-19-823994-7].

Grayling, A.C. (2005), *Descartes: The Life and Times of a Genius*, Walker Publishing Co., New York [ISBN 0-8027-1501-X].

Watson, Richard A (2007), *Cogito, Ergo Sum: a life of Rene Descartes*, David R Godine USA [ISBN 978-1-56792-335-3]. (This book was chosen by the New York Public library as one of "25 Books to Remember from 2002".)

3. See, for example, Shorto, Russell (2008), *Descartes' Bones*, Doubleday, USA [ASIN: B005UG265Q].

4. For more on Fermat and his theorem, see:
Mahoney, Michael Sean (1994), *The mathematical career of Pierre de Fermat, 1601–1665*, Princeton University Press, Princeton [ISBN 0-691-03666-7].
Singh, Simon (2002), *Fermat's Last Theorem*, Fourth Estate Ltd. UK [ISBN 1-84115-791-0].
Simmons, George F. (2007), *Calculus Gems: Brief Lives and Memorable Mathematics*, Mathematical Association of America USA [ISBN 0-88385-561-5].

Figure Credits

5. Figure 17.1 courtesy: Frans Hals, via Wikimedia Commons (from public domain). https://commons.wikimedia.org/wiki/File:Frans_Hals_-_Portret_van_Ren%C3%A9_Descartes.jpg

18

Life's Infinite Variety : Finding Order in Species

Shakespeare's *Antony and Cleopatra* depicts a character, Lepidus, who, in a drunken mood, proclaims, "Your serpent of Egypt is bred now of your mud by the operation of your Sun; so is your crocodile" (Act II, Scene VII). It is unlikely that either Shakespeare or the Romans believed that snakes and crocodiles came out of mud spontaneously; they had definitely seen eggs being laid and hatched. However, people — even learned ones — were not so sure, in the sixteenth century, about smaller insects. The prevailing idea indeed was that the small creatures like worms and vermin did emerge from filth and mud spontaneously.

Such a view led to two different ways life could be created. Every ancient civilization knew how horses or cattle, say, produced their offspring. Many of them also knew how to selectively breed these animals to enhance specific qualities. It was never suggested that horses came out of mud in the farmland. But ancient thinkers, including Aristotle, assumed that the situation was different when it came to much smaller creatures.

The reason for this dichotomy is not difficult to understand. It was almost an everyday experience for the common man to see maggots appearing in rotting meat. But nobody had ever seen a horse appearing out of mud. This led to the doctrine of 'spontaneous generation', which claimed that certain kinds of creatures could come up spontaneously.

Doubts about this opinion first arose in the seventeenth century, following the work of Francesco Redi (1626–1697), a physician (and, incidentally, a poet whose Bacchanalian poem in praise of Tuscan wines is still read in Italy today!) from

T. Padmanabhan and V. Padmanabhan, *The Dawn of Science*,
https://doi.org/10.1007/978-3-030-17509-2_18

Tuscany, Italy [1]. Around the time Redi was born, William Harvey had already published a book in which he had suggested that small creatures probably came from eggs too tiny to be seen. Redi read it and came up with a simple experiment to settle this issue.

Redi knew that decaying meat not only produced flies but also attracted them in large numbers. It occurred to him that a first generation of flies could be laying eggs on the meat from which the second generation originated. He prepared eight different flasks with a variety of meat placed in them. He sealed four of them totally airtight and left the other four open. The flies could now land only on meat in the open flasks and indeed it was found that only the four open flasks produced maggots. Meat in the closed flasks was as putrid and smelly as the others, but there were no living creatures.

Redi also repeated the experiment by not sealing the flasks but covering them with gauze, which let in the air but not the flies. And again no life form developed in the insulated meat. This was probably the first biological experiment using a controlled sample. These results were published in 1668 as *Esperienze Intorno alla Generazione degli Insetti* (Experiments on the Generation of Insects). This book is regarded as his masterpiece and a milestone in the history of modern science.

Knowing full well what happened to other outspoken thinkers (like Giordano Bruno and Galileo Galilei), Redi was careful in expressing his new views in such a manner that they would not contradict theological traditions of the Church. His interpretations were always defended by biblical passages, such as his famous quote: *omne vivum ex vivo* (All life comes from life)!

Redi's experiment strongly suggested that flies originated from other flies and not spontaneously from decaying meat. However, this did not convince people to give up the older idea. It is said that Redi himself believed that there could be very small creatures which actually came into life spontaneously even though flies were not born this way! This idea, interestingly enough, had gathered support from an unexpected quarter — viz., the advent of microscopy, allowing one to see very tiny creatures.

Galileo, when he was using the telescope, realized that there could also be an arrangement of lenses which would magnify small objects. The actual theory of microscopes was developed by Kepler and Torricelli; by the mid-seventeenth century, several investigators had started using microscopes to study biological specimens.

The first person to study biological specimens systematically using microscopes was probably Marcello Malpighi (1628–1694), an Italian physician who lectured in several Italian universities, and especially in Bologna [2]. He began his biological

Fig. 18.1: Portrait of Francesco Redi (1626–1697) from Tuscany, Italy, who raised serious doubts about the spontaneous generation of life [8]. He was probably the first to perform a biological experiment using a controlled sample to demonstrate a result. Redi, who was a physician by profession, was also a successful poet. One of his, rather Bacchanalian, poems in praise of Tuscan wines, is still read in Italy today!

investigations by studying the lungs of frogs and almost figured out the entire process of respiration. He also discovered very fine blood vessels allowing the transfer of blood from arteries to veins, which conveniently completed the theory of blood circulation.

The most famous scientist to use the microscope was, however, Antony van Leeuwenhoek (1632–1723) who hailed from Delft in Netherlands. Making microscopes became Leeuwenhoek's passion, and he used his lenses to observe virtually everything around him [3]. Being a master lensmaker (Fig. 18.2), he could make a single-lens microscope with virtually no imperfections so that it could magnify objects up to 200 times. Unfortunately, none of Leeuwenhoek's microscopes have

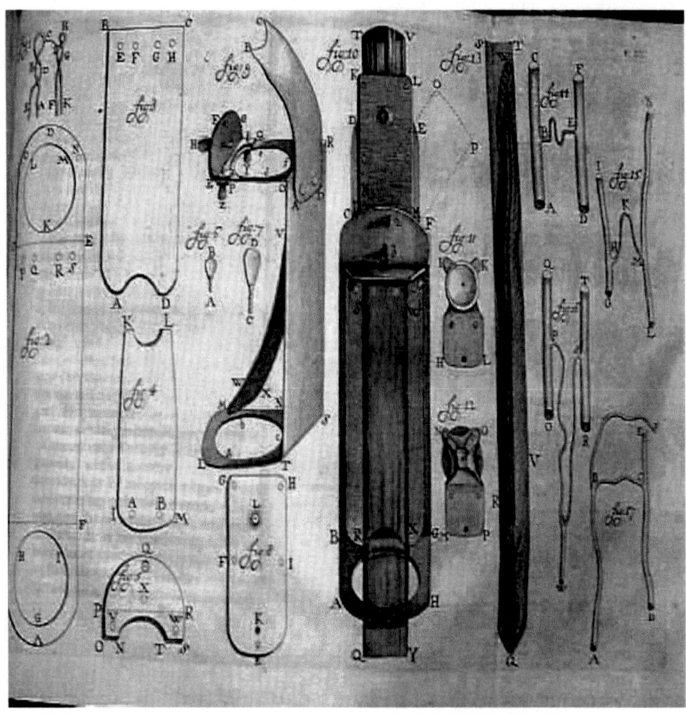

Fig. 18.2: Leeuwenhoek's microscope and sketches with some details of its construction [9]. The idea of a microscope was known to several scientists including Galileo, Kepler, and Torricelli. In fact, by the mid-seventeenth century, there were several investigators who were using microscopes to study biological specimens. Leeuwenhoek (1632–1723) was the most famous — and also probably the most successful — amongst them. He was an expert in making perfect lenses, and his microscopes, usually incorporating a single lens, could magnify objects nearly 200 times.

survived, so no one really knows how exactly he managed to grind such a perfect lens.

By the end of the seventeenth century, Leeuwenhoek had a virtual monopoly on this research. So much so that his contemporary Robert Hooke (1635–1703), another microscope enthusiast, bemoaned the way the field had come to rest entirely on one man's shoulders! Over the years, several prominent individuals, including the Russian Tsar Peter the Great, visited Leeuwenhoek. It was claimed that Leeuwenhoek refused to reveal the really powerful microscopes he relied on

Fig. 18.3: Left: Portrait [10] of Marcello Malpighi (1628–1694). Right: Portrait [11] of Lazzaro Spallanzani (1729–1799). Marcello Malpighi was an Italian physician who lectured in Bologna. His investigations of biological specimens using microscopes led to the discovery of — what we now call — capillaries, in which blood flows from arteries to veins. This was crucial in completing the understanding of blood circulation. Both he and Lazzaro Spallanzani were skeptical of the spontaneous generation of life and performed careful experiments to demolish this idea, but in vain.

for his discoveries, and instead showed his visitors a collection of rather mediocre apparatus!

In 1675, Leeuwenhoek discovered the existence of 'living things' in ordinary ditch water which were too small to be seen by the naked eye. These 'animalcules' — which are now called protozoa — were as alive as an elephant or a man. He found that the common yeast used for ages in bread-making was actually an association of tiny living creatures much smaller even than the 'animalcules'. In 1683, Leeuwenhoek observed still smaller organisms, which we now call 'bacteria'. In fact, no one else was able to observe bacteria again with the same clarity for over a century, until compound microscopes were devised, able to magnify as much as the tiny lenses of Leeuwenhoek had done!

Around this time, Robert Hooke (1635–1703), a contemporary of Leeuwenhoek, made an equally important discovery — that of the cell — using a microscope.

Fig. 18.4: Portrait [12] of Antony van Leeuwenhoek (1632–1723), whose experiments with microscopy led him to discover, in 1675, the existence of 'living things' in ordinary ditch water. These 'animalcules', as he called them — now known as protozoa — were as alive as any other large life form. In 1683, Leeuwenhoek observed still smaller organisms, which we now call 'bacteria'. Rather ironically, these observations seemed to support the notion of spontaneous generation of life. This is one of the (fortunately, rare) examples of better observations ending up supporting the wrong scientific theory!

This turned out to be of great significance as biology developed. In 1661, King Charles II of England had commissioned Sir Christopher Wren to create a set of pictures based on microscopic studies. Wren started in earnest but soon found that he didn't have the time to do it. So he handed the project over to an up-and-coming scientist with good drawing skills, viz., Robert Hooke — who was 26 years old when he took on this assignment.

Hooke, who was already fascinated by the world opened up by microscopes, exceeded the King's commission and did his own bit of microscopic observation of various objects. He published them in 1665, in the book *Micrographia*, which contained some of the most beautiful drawings of his observations. Among them was the result of looking at a thin slice of cork. Hooke noted that the slice was made of a fine grid of tiny rectangular chambers which he called 'cells'. (He said that they resembled the cellula, small rooms which monks inhabited.) Later developments in biology proved the importance of this discovery.

Following the discovery of cells, Leeuwenhoek also observed the vacuole of the cell and the spermatozoa. Leeuwenhoek had also figured out how to make a broth (by soaking pepper in water) in which protozoa could multiply. In fact, such broths appeared to produce protozoa on their own. Even when a broth was first boiled and filtered, to eliminate any sign of protozoa, it showed signs of these organisms within a short time thereafter. The tiny living organisms Leeuwenhoek found actually reinforced people's belief in spontaneous generation. Many took this to be a clear sign of the generation of life from non-life.

There were, fortunately, a few skeptics [4]. One among them, Lazzaro Spallanzani (1729–1799), had the insight to repeat the broth experiment by sealing off the neck of the flask which contained the broth. He found that the broth did not develop microscopic life; but the adherents of spontaneous generation maintained that the process had somehow removed the 'vital spirit' from the broth. The issue was not settled until much later, at the time of Louis Pasteur (1822–1895).

Box 18.1: Microscopy is Alive and Well!

Once microscopes were being marketed *en masse*, over a period of time, several people contributed to refining the details so as to make them more powerful and easier to use. In addition to the improved mechanical design, this also involved using various properties of light — and more generally — electromagnetic radiation. We have microscopes which make use of ultraviolet light, fluorescence, polarization of radiation, and manipulation of radiation frequency — just to mention a few.

In the 1870s, Ernst Abbe developed a theory about the ultimate limitation one would face in the magnification produced by *any* microscope using light. This limitation, it was thought, arises from the fact that light is a wave. So it was believed that when you probe smaller and smaller sizes, you will eventually reach a stage at which the fuzziness associated with the wave nature of light will be the limiting factor. (This is closely related to a phenomenon called diffraction.) This theoretical consideration suggested that we would not be able to see structures smaller than about 200 nanometres (which is roughly 200 times smaller than the width of a hair).

Interestingly enough, the Romanian chemist Stefan Hell found a way to circumvent this barrier, which people had believed in for nearly a century! He did this by using a sophisticated technique that combined fluorescence with a sweeping light. Hell and his colleagues could directly probe the insides of a living cell using this technique, something even the electron microscope could not do. This got Hell the 2014 Nobel Prize for Chemistry.

So Leeuwenhoek's legacy lives on!

Meanwhile, biological understanding was growing on another front, viz., understanding the variety of living species. The ancient civilizations knew of only a few hundred species of living beings. Aristotle, probably the keenest observer among the Greeks, could list only about 500 species of animals, and his student, the ancient Greek botanist, Theophrastus, knew of only about 500 species of plants. In medieval days, several naturalists attempted to extend the list of animal species and produce a systematic classification of them.

The problem of classification of animals is far more difficult than one would imagine at first. Consider, for example, the question of defining a 'bird'. Calling it a 'two-legged' creature will, of course, make man a bird, while calling it a 'winged creature' will make the bat a bird. (Incidentally, an eighteenth century naturalist once told Voltaire that the briefest definition for a man is a 'featherless biped' to which Voltaire retorted that biologists could make a man out of a plucked chicken!)

Given such perplexity, it was not surprising that the English biologist, John Ray (1627–1705), had to spend nearly a whole lifetime in classifying plants and animals [5]. In 1667, after many years of painstaking travels and observations, he published a catalogue of plants in the British Isles. Later, during 1686 to 1704, he enlarged these studies and published a three-volume encyclopedia describing over 18 600 plants. He also published, in 1693, the first logical classification of animals based on their hooves, toes, and teeth.

This classification held sway for a long time, until the Swedish naturalist, Carl Linnaeus (1707–1778), produced a far more detailed, scientific classification [6]. This 'binomial system', used worldwide today, consists of Latinized two-part names assigned to every species. The first part, the genus, is typically the name for a small group of closely related organisms. The second part, which is the specific epithet, is used to identify and distinguish a particular species from others belonging to the same genus. Together, they give the full scientific name for an organism.

Linnaeus introduced this in his book, *Systema Naturae*, which grew from 12 outsize pages in its first edition in 1735, to 2400 pages in its twelfth edition, in a

Fig. 18.5: Portraits of John Ray (left) [13] and Carl Linnaeus (right) [14]. John Ray (1627–1705) spent nearly his whole lifetime classifying plants and animals. In 1667, he published a catalogue of plants of the British Isles which was developed into a three volume encyclopedia describing over 18 600 plants. He also came up with the first logical classification of animals, which was the best available for a long time, until the Swedish naturalist, Linnaeus (1707–1778), produced a more detailed one. Linnaeus was a professor at the Uppsala University and was extremely popular with the students. His lectures used to be held in the Botanical Garden and his botanical excursions, made every Saturday during summer, where he and his students explored the flora and fauna in the vicinity of Uppsala, was an added attraction for the students. He had numerous students (called his 'apostles') who carried out the meticulous job of collecting, classifying, and cataloguing the specimens. Incidentally, it is believed that Linnaeus invented the card index [7] as an effective cataloguing method, for quick reference and retrieval of information about his huge collection of thousands of species!

span of 30 years. Through this huge effort, Linnaeus succeeded in organizing the chaotic natural world. Incidentally, he was totally against the idea of evolution and insisted that all species were created separately once and for all. So it is ironic that the seminal work by Charles Darwin and Alfred Russel Wallace on the theory of evolution appeared about a century later, in 1858, in the *Journal of the Linnaean Society*, named after Linnaeus!

The greater understanding of the animal kingdom acquired by these investigations paved the way for tackling much wider issues. The schemes of classification clearly showed two features: the number of species which inhabit the Earth was much greater than what the ancient wise men like Aristotle had imagined, and there were some patterns of similarity between animals of very different kinds when viewed from a fundamental basis. (One familiar example, known to common man, is between the domestic cat and the tiger.) Very soon people started wondering where all these different species came from. Indeed, if different species produced only offsprings of their own kind, it logically followed that all the species must have existed from the very beginning. On the other hand, if species did 'change', it must have been at a very slow rate and it would have taken an incredibly long time to produce the kind of variety which we see. Questions of this nature were the ones that led Darwin to his remarkable discovery years later.

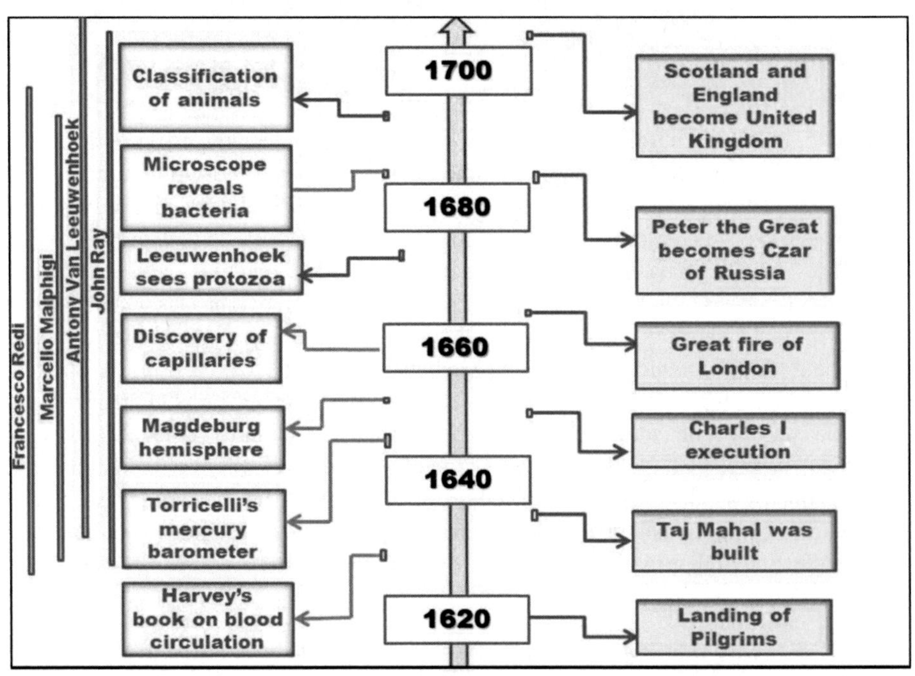

WHEN

Notes, References, and Credits

Notes and References

1. See, e.g., Hawgood B.J. (2003), *Francesco Redi (1626–1697): Tuscan philosopher, physician and poet*, Journal of Medical Biography **11** (1), 28–34 [doi:10.1177/096777200301100108].

2. See, e.g., Daston, Lorraine (2011), *Histories of Scientific Observation*, University of Chicago Press, Chicago [ISBN 978-0226136783].

3. For more details, see:
   Chung, King-thom and Liu, Jong-kang (2017), *Pioneers in Microbiology: The Human Side of Science*, World Scientific Publishing, Singapore [ISBN 978-9813202948].
   Schierbeek, A. (ed) (1959), *Collected Letters of A. van Leeuwenhoek, Measuring the Invisible World*, Abelard-Schuman, London and New York. (This book contains excerpts of van Leeuwenhoek's letters, focusing on his priority in several new branches of science, and also makes several important references to his spiritual life and motivations.)

4. See, e.g., Kruif, Paul de (2002), *Microbe Hunters*, Mariner Books, USA [ISBN 978-0-15-602777-9].

5. For more on John Ray, see, e.g.,
   Mandelbrote, Scott (2004), *Ray, John (1627–1705), naturalist and theologian*, Oxford Dictionary of National Biography, Oxford University Press, Oxford [ISBN 9780198614128] [doi:10.1093/ref:odnb/23203].
   Raven, Charles E. (1950), *John Ray, naturalist: his life and works* (2nd edn.), Cambridge University Press, Cambridge [ISBN 9780521310833].

6. See, e.g., Lars Hansen (ed) (2007–2011), *The Linnaeus Apostles — Global Science and Adventure* (8 vols), The IK Foundation and Company, London and Whitby [ISBN 978-1-904145-26-4].
   Greene, Edward Lee (1912), *Carolus Linnaeus*, University of California Libraries USA [ASIN: B007Q4FKDU].

7. British Society for the History of Science, *Carl Linnaeus Invented the Index Card*, ScienceDaily, 16 June 2009.
   www.sciencedaily.com/releases/2009/06/090616080137.htm.
   Muller-Wille, Staffan and Scharf, Sara (2009), *Indexing Nature: Carl Linnaeus (1707–1778) and his Fact-Gathering Strategies*, University of Exeter & University of Toronto, Toronto.

Figure Credits

8. Figure 18.1 courtesy: Painting by Jacob Ferdinand Voet, via Wikimedia Commons. https://commons.wikimedia.org/wiki/File:Francesco_Redi,_founder_of_experimental_biology.jpg (from public domain).
9. Figure 18.2 courtesy: Henry Baker [Public domain], via Wikimedia Commons. https://commons.wikimedia.org/wiki/File:Van_Leeuwenhoek%27s_microscopes_by_Henry_Baker.jpg (from public domain).
10. Figure 18.3 [Left] courtesy: Carlo Cignani [Public domain], via Wikimedia Commons. https://commons.wikimedia.org/wiki/File:Marcello_Malpighi_by_Carlo_Cignani.jpg (from public domain).
11. Figure 18.3 [Right] courtesy: : Lazzaro Spallanzani [Public domain], via Wikimedia Commons. https://commons.wikimedia.org/wiki/File:Spallanzani.jpg (from public domain).
12. Figure 18.4 courtesy: Jan Verkolje [Public domain], via Wikimedia Commons. https://en.wikipedia.org/wiki/File:Anthonie_van_Leeuwenhoek_(1632-1723)._Natuurkundige_te_Delft_Rijksmuseum_SK-A-957.jpeg (from public domain).
13. Figure 18.5 [Left] courtesy: Magnus Manske on en.wikipedia [Public Domain] via Wikimedia Commons. https://commons.wikimedia.org/wiki/File:John_Ray.jpg (from public domain).
14. Figure 18.5 [Right] courtesy: Alexander Roslin [Public domain], via Wikimedia Commons. https://commons.wikimedia.org/wiki/File:Carl_von_Linn%C3%A9.jpg (from public domain).

19

A Measure of the Heavens

After the Copernican Revolution, physics — or 'natural philosophy' as it was then called — was progressing rapidly with contributions from Descartes, Torricelli, Boyle, Pascal, and many others. Some of the important contributions to physics and astronomy in the later half of the seventeenth century came from Christiaan Huygens (1629–1695) from the Netherlands [1]. Unfortunately, the importance of his work was overshadowed by the towering dominance of Newton's work, which was contemporaneous.

Born into a wealthy family in the Hague, Huygens had the advantage of an excellent education. Amongst his father's friends were several European intellectuals with whom he corresponded on a regular basis. In 1645, Huygens joined the University of Leiden, where he studied law and mathematics. His first breakthrough came in 1655 somewhat accidentally. While helping his brother to make a telescope, Huygens stumbled upon a new and better method for grinding lenses, which enabled him to build telescopes with much higher resolutions. He promptly built an astronomical telescope nearly 23 feet (7 meters) long, started scanning the sky, and came up with several important discoveries.

The first was the existence of a huge cloud of dust and gas in the constellation of Orion, which we now call the Orion nebula. Huygens also found a satellite of Saturn, which was as big as any of the Jovian satellites, and named it 'Titan'. (In Greek mythology, Saturn was indeed the leader of a group of gods called the Titans.) With this discovery, the Solar System ended up having an equal number of planets and satellites, viz., six each. Huygens then fell into a natural trap and

T. Padmanabhan and V. Padmanabhan, *The Dawn of Science*,
https://doi.org/10.1007/978-3-030-17509-2_19

Fig. 19.1: Portrait of Christiaan Huygens (1629–1695), who was the first to construct a very powerful astronomical telescope, a micrometer, and an accurate pendulum clock [3]. His telescope was about 23 feet (7 meters) long, and using it to observe the skies, he discovered the Orion nebula (as we call it today), the largest satellite of Saturn (which he named Titan), and the detailed structure of Saturn's rings.

declared to the world that no more satellites or planets remained to be discovered; but within his lifetime, four more satellites were observed! Huygens' telescope also revealed the detailed structure of a ring that surrounded Saturn without touching the planet. (Galileo had noticed the ring-like structure around Saturn earlier, but Huygens' more powerful telescope revealed its details.) These rings were unique and attracted quite a lot of popular attention.

Huygens also tried to make quantitative estimates in many areas of astronomy. He was the first person to make a serious effort to determine the distances to the stars. Assuming that the brightest star in the sky, Sirius, was as bright as the Sun, he estimated its distance to be 2.5 trillion miles. (The actual distance turns out to be 20 times larger, because Sirius is actually much brighter than the Sun. One shouldn't be too harsh on Huygens for this. Errors of similar nature — and for similar reasons — are routinely made in astronomy!) While engaged in making these quantitative measurements, Huygens realized the need for accurate measuring devices, and he developed two such instruments. The first was a micrometer which could measure angular separations as small as a few arcseconds. The second was a pendulum clock, which brought Huygens well-deserved fame and glory.

The best clocks available in Huygens' day were those designed in the Middle Ages. These were complicated mechanical devices powered essentially by falling weights. Though such apparatus was sufficient to produce fancy-looking decorative pieces for royal courts, their accuracy was limited to about a fraction of an hour; this made them totally useless for scientific studies. To build a more accurate clock, one needed a device which maintained a constant period during the motion. It had been known since Galileo's time that a pendulum did this fairly accurately. All that was required was to connect the pendulum to suitable gear wheels and attach falling weights to supply the energy lost due to friction. This was essentially what Huygens did and thus the first 'grandfather clock' was born. To improve its accuracy, Huygens added one crucial refinement. Since he knew that the pendulum kept a constant period only approximately, he adjusted the movement in such a way that the period would remain *exactly* constant. With this refinement, his clock was accurate to a fraction of a minute (see Fig. 19.2).

While Huygens contributed extensively to different areas, one cannot but think that he had a particular fascination for time-keeping devices. For decades after the initial fabrication of the pendulum clock, all Huygens' efforts went into measuring the flow of time by a reliable mechanism, which would be immune to disturbance from the environment. For example, he worked on the design of a marine clock which could be used on board ships. All these required careful theoretical considerations and exquisite design skills. In fact, some of his explorations required the use of abstract mechanics, applying the latest developments in mathematics; he even needed to develop a new branch of mathematics, called the theory of evolutes.

The pendulum clock and its descendants helped to spread Huygens' name throughout Europe. He was elected a member of the Royal Society in 1663 and was invited to the Paris court of Louis XIV in 1666. He did spend some time in Paris, but returned to the Netherlands in 1681 because Louis XIV started promulgating ordinances against Protestants — and Huygens was a Protestant.

Interestingly enough, Huygens' most significant contribution did not get any recognition during his lifetime. This related to the nature and propagation of light. Huygens firmly believed that light, especially its propagation, could be understood

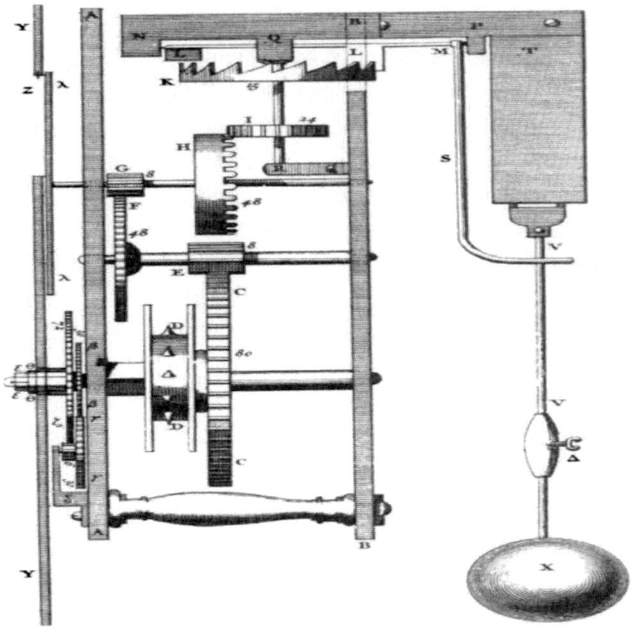

Fig. 19.2: Huygens' design of the pendulum clock [4]. The best clocks available in those days, designed in the Middle Ages, were good enough only as decorative pieces in royal courts because their accuracy was limited to about a fraction of an hour. It was known, from Galileo's time, that a pendulum maintained reasonably accurate periodic motion. Huygens connected the pendulum to suitable gear wheels and attached falling weights to supply the energy lost due to friction. He also improved its accuracy, by adjusting its movement in such a way that the period would remain *exactly* constant. Huygens' clock was accurate to within a fraction of a minute!

as a wave phenomenon. But this idea was in violent conflict with the prevailing notion that light rays were corpuscular in nature (that is, made of tiny particles). The strongest evidence against Huygens' ideas came, in fact, from everyday experience — that light always travelled in a straight line. This is to be contrasted with typical wave propagation like that of sound; while sound waves can 'bend around' obstacles, light cannot. (That is why we can hear someone speak from around a corner even when we can't see that person.) Huygens showed that there were indeed specific conditions under which light may appear to travel in a straight line even though it was fundamentally of wave nature. Unfortunately, no one took his ideas seriously, to a large extent due to the dominating influence of Newton's concepts, which assumed that light was made up of particles. And so Huygens' (correct) ideas about light lay buried for another century.

Box 19.1: From Mathematics to Marketplace

Scientists are usually considered to be not very street-smart, and in fact rather unconcerned about the commercial side of their work. Not Huygens! His idea was to carefully protect the possible fortune that might result from any of his inventions.

For example, when he invented his first marine clock — which was essentially a portable clock that was accurate enough to be used to determine the time and longitude at sea — he postponed the publication of its details. This was because it appeared that his design would do well during the trials at sea and could evolve into a very practical (and profitable) gadget. A publication would have interfered with the exclusive license Huygens had given to a Hague clockmaker, and the success from rigorous tests at sea would have increased the clock's commercial value. For nearly two years in the mid-1660s, Huygens strived to secure patents from the French, English, and Dutch governments, and was competing for the reward offered by the Spanish throne for a reliable marine device. All this, while his scientist colleagues were eagerly awaiting the promised treatise!

Huygens' disputes with the clock-makers Douw and Thuret over patent rights — though not exactly unprovoked — were more acrimonious than any of his debates with other scientists over issues which had no commercial consequence. Huygens knew what his inventions were worth and was not giving anything away through any kind of other-worldly attitude.

Meanwhile, around the same time [2], astronomical discoveries similar to those of Huygens were also being made by the French astronomer Giovanni Cassini (1625–1712). (Cassini was born Italian but became a naturalised French citizen.) Cassini started with a detailed study of Jupiter's moons and measured Jupiter's period of rotation. He then turned his attention to Saturn and discovered four new satellites, which were named Lapetus, Rhea, Dione, and Tethys. In this process, he disproved Huygens' conjecture that there must be an 'equal number of satellites

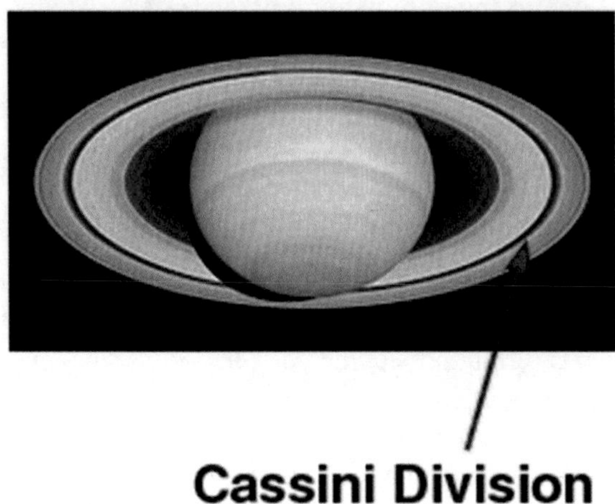

Fig. 19.3: The French astronomer Giovanni Cassini (1625–1712) made astronomical discoveries similar to those of Huygens and at around the same time. He discovered four new satellites of Saturn — Lapetus, Rhea, Dione, and Tethys — and also observed that the ring around Saturn was double; the gap between the rings is called the Cassini division in his honour [5].

and planets in the Solar System'. He also discovered that the ring around Saturn was double; the gap between the rings is named after him and is called the Cassini division (see Fig. 19.3).

Cassini's most valuable contribution, however, was the measurement of the distance to Mars. By comparing his own observations of the position of Mars with those made by the French astronomer, Jean Richer (1630–1696), he could determine the distance to Mars. The *relative* distances of the Sun and the planets had been known quite accurately since the days of Kepler. Therefore, knowing any one distance, all the other distances could be determined. In particular, Cassini computed the distance between the Earth and the Sun to be about 87 million miles which is only about 7 per cent lower than the actual distance. In fact, this was the first measurement to give a value that was even approximately correct; previous estimates had been in the range of five to 15 million miles.

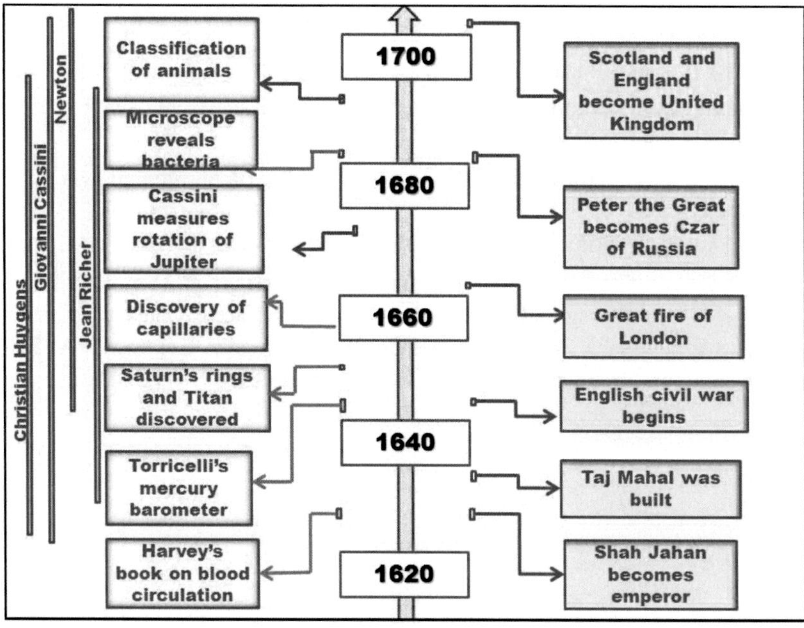

WHEN

Notes, References, and Credits

Notes and References

1. For more on Huygens, see, e.g.,
Garber, Daniel (2003), *The Cambridge History of Seventeenth-Century Philosophy* (2 vols.), Cambridge University Press, Cambridge [ISBN 978-0-521-53720-9].
Andriesse, C.D., (2005), *Huygens: The Man Behind the Principle*, Cambridge University Press, Cambridge [ISBN 978-0521850902].
Hooijmaijers, H. (2005), *Telling time – Devices for time measurement in Museum Boerhaave – A Descriptive Catalogue*, Museum Boerhaave, Leiden [ISBN 9062920004].
Taylor, John C. and Kersen, Frits Van ((2004), *Huygens's Legacy — The golden age of the pendulum clock*, Fromanteel Ltd, Castle Town, Isle of Man [ISBN 978-0954833909].
2. See, e.g., Rene, Taton (2008), 'Cassini, Gian Domenico' in *Complete Dictionary of Scientific Biography*, Charles Scribner's Sons, Detroit, pp. 100–104, [ISBN 9780684315591].

Figure Credits

3. Figure 19.1 courtesy: Unknown, 17th century artist. [Public domain] via Wikimedia Commons. https://commons.wikimedia.org/wiki/File:Christiaan_Huygens.gif (from public domain).
4. Figure 19.2 courtesy: Christiaan Huygens (1629 – 1695) [Public domain] via Wikimedia Commons. https://commons.wikimedia.org/wiki/File:Huygens_clock.png (from public domain).
5. Figure 19.3: The image of Saturn is courtesy NASA, ESA and E. Karkoschka (University of Arizona); Public domain. [Material was created for NASA by Space Telescope Science Institute under Contract NAS5-26555]. https://commons.wikimedia.org/wiki/File:Saturn_HST_2004-03-22.jpg.

20

Calculus Developed in South India

An ancient text from Kerala, a southern state in India, contains a verse giving the circumference of a circle. The translation goes as:

"Multiply the diameter by four. Subtract from it and add to it alternately the quotients obtained by dividing four times the diameter to the odd numbers 3, 5, etc."

In modern notation, this leads to a remarkable infinite series expansion for $\pi/4$ given by:

$$\frac{\pi}{4} = 1 - \frac{1}{3} + \frac{1}{5} - \frac{1}{7} + \cdots .$$

This result — usually attributed in textbooks to Gregory (1638–1675) and others, who were born more than a century after the text in question was authored — is one of the many gems to be found in the ancient Kerala texts (especially the one called *Yukti-bhasha*), which laid the foundation for a branch of mathematics that we now call calculus [1], centuries before Newton and Leibnitz.

The discovery of the contributions from the Kerala school of mathematicians is rather recent and many details are still being probed extensively. *This is in spite of the fact that some of these ancient Kerala texts were referred to explicitly in an article by Charles M. Whish as early as 1834.* Whish — having learnt the local language, Malayalam, and collected palm leaf manuscripts from Kerala — found to his astonishment a "complete system of fluxions" in them. He published [2] a paper in 1834 in the Transactions of the Royal Asiatic Society of Great Britain and Ireland with an explicit title "On the Hindu quadrature of the circle and the infinite series

T. Padmanabhan and V. Padmanabhan, *The Dawn of Science*,
https://doi.org/10.1007/978-3-030-17509-2_20

of the proportion of the circumference to the diameter exhibited in the four *sastras*, the *Tantra-sangraham, Yukti-bhasha, Carana-Padhati*, and *Sad-ratna-mala*."

In addition to the result for π quoted above, these texts contain the infinite series expansion for $\tan^{-1}(x)$ (from which the above result is but one step; you just put $x = 1$), the series expansion for sine $(\sin)$ and cosine $(\cos)$, and the *indefinite integral* of x^n — just to list a few. Virtually all these results were obtained using techniques which, in the current terminology, belong to the branch of mathematics that we call calculus.

Who were these mathematicians and how and why did they develop these techniques? Though many of the details are still sketchy, historians of mathematics have now put together the following picture [3].

Most of these developments took place in villages along a river called Nila in the ancient days (and currently called the river Bharatha, the second longest river in Kerala), from 1300–1600 AD. One of the main villages was called Sangama-grama in ancient times; it is thought to refer to the village called Irinhalakkuta (about 50 km to the south of Nila) in present day Kerala. There are a few other towns, like Kudalur and Tirunavaya, for which one could raise arguments in favour of their being Sangama-grama, so this issue is not yet completely settled.

What is certain is the existence of an extraordinary lineage of mathematicians in the Sangama-grama, of which Madhava ($\sim$ 1350–1420) seems to be the one who developed many of the basic ideas of calculus. The infinite series for the sine, cosine, and arctan, as well as the rudiments of integration, are attributed to Madhava by many later sources. He was strongly influenced, like many other Indian scholars of that period, by Aryabhata (476–550 AD) — which is no surprise, since *Aryabhateeyam* (circa 500 AD) was indeed a very influential text in India, as it was (through its translations) in the Arab world and medieval Europe.

Box 20.1: How do you 'Develop' Calculus?

The essence of calculus involves the concept of an 'infinitesimal' — something which you can take to be as close to zero as you please and yet manipulate it systematically in a manner which you cannot quite do with zero. This is the concept used by Archimedes in a primitive form, developed (and used) systematically by Madhava in the South Indian school of mathematics, then developed in more generality by Leibnitz and Newton later on. Though the concept is rather technical — whence this chapter is necessarily more technical than the rest — we will try to convey a flavour of it for the non-expert reader.

Think of a car moving along a straight road with *varying* speed. If you are driving the car, you know that at any given instant of time, it has a well-defined speed. We would like to define this notion of 'speed at a given instant' by a sensible mathematical procedure. One natural way of doing this is as follows:

Suppose you want to know the speed at a given time t. Take two instants of time close to t, say, $t - \epsilon$ and $t + \epsilon$, where ϵ is a small number, say, one minute. Suppose the car travelled a distance x between these two instants of time. Then the *average* speed of the car during this interval — around the instant t — will be $x/2\epsilon$.

But this only gives you the *average* speed of the car during the interval $(t - \epsilon)$ to $(t + \epsilon)$ which may not be the same as the *instantaneous* speed at the time t. To tackle this issue, you can try to make ϵ smaller and smaller so that you are computing the average very, very close to the instant t at which you want to know the speed of the car. But as ϵ gets smaller and smaller, the distance x traveled by the car in the small interval 2ϵ also gets smaller and smaller. So the average speed $x/2\epsilon$ will be a ratio of two very small numbers, one in the numerator and one in the denominator. Ideally, we want to take ϵ to be zero; but then x will also be zero and the average speed will reduce to the meaningless expression $0/0$.

One of the main triumphs of calculus is to give meaning to such expressions when both x and ϵ tend to zero. In this limit, we say that both x and ϵ are *infinitesimals*, while their ratio is a finite number — which is the instantaneous speed indicated by the speedometer of the car. The infinitesimals, which can be taken to be as close to zero as we want, help to give meaning to ideas like the instantaneous speed of a car. This entire process is known as differentiation.

The word used for the infinitesimal by Madhava, the Indian mathematician who developed these ideas in the fourteenth century, is 'sunya-prayam'. It is a Malayalam word meaning 'like zero', 'of the size of zero', etc. The fact that you can make the interval as close to zero as you like is captured by the use of the word 'yadhestam', which roughly translates to 'as you desire'. Once Madhava had the concept of infinitesimal, he could build the rest of the concepts in his own tradition.

The reverse process is called integration, in which you put together an infinite number of infinitesimal quantities to make a finite quantity. This is the idea which Archimedes used implicitly to obtain the area and volume of a sphere. Madhava also developed the concept of integration in his tradition of mathematics. Centuries later, the same ideas were developed in greater generality by Leibnitz and Newton.

Madhava founded a lineage which probably lasted till the late seventeenth century. His student Parameshwara ($\sim$ 1360–1455) was an astronomer–mathematician, who authored more than two dozen works and made astronomical observations for an extended period of nearly five decades. His son Dhamodara was the teacher of another key figure, Neelakanta Somayaji ($\sim$ 1450–1550), the author of *Tantra-sangraha*, which contains extensive discussions of astronomy and related mathematics. Yet another student of Dhamodara was Jyeshtadeva ($\sim$ 1500–1610), the author of *Yukti-bhasha*, which probably contains the clearest exposition of some of the topics in calculus.

The quotation at the beginning of this chapter is in *Yukti-bhasha*, which attributes it to *Tantra-sangraha* which, in turn, attributes the result to Madhava. Similarly, there is another detailed verse in *Yukti-bhasha* which attributes to Madhava the infinite series expansion for the inverse tangent. The first sentence of *Yukti-bhasha* states that it is going to describe in detail all the mathematics "useful in the motion of heavenly bodies following *Tantra-sangraha*"; but actually it achieves much more as an independent treatise than merely functioning as a commentary. (It is also remarkable for being written in Malayalam — not Sanskrit, which was the standard language for learned expositions — and is in prose, not poetry.) Unlike *Tantra-sangraha*, this work provides detailed arguments for the results and is written in a style which is very straightforward and unpretentious. (An English translation with detailed commentary is now available; see [1].)

Here, we will concentrate on two specific results in the *Yukti-bhasha* and highlight the role played in it by calculus, as we call it today. The discussion in what follows, *has to be* more technical than in other chapters of this book to demonstrate the origin of calculus; so we shall not shy away from some technical details.

Let us start with the series expansion for $\pi/4$. To understand what is involved in deriving the infinite series expansion for $\pi/4$ given at the beginning of this chapter, it is useful to recall the procedure in modern notation. If $t = \tan\theta$ with $0 \le \theta \le \pi/4$, then we have $\mathrm{d}t/\mathrm{d}\theta = 1 + t^2$. Rewriting this equation by taking the reciprocals (which is actually a rather clever step!) and using the geometric series expansion, one gets $\mathrm{d}\theta/\mathrm{d}t = 1 - t^2 + t^4 + \cdots$. Integrating both sides of this equation between 0 and 1 will give $\tan^{-1}(1) = \pi/4$ as an infinite series. (The *indefinite* integral also gives the infinite series expansion of θ in terms of $\tan\theta$.) Clearly, the derivation involves the concepts of differentiation and integration in some form, as well as knowledge of the indefinite integrals of powers. How did the ancient Kerala mathematicians do this?

The idea was to consider the geometry of the circle and give meaning to all the relevant quantities in terms of suitably defined geometrical constructions. In Fig. 20.1, AB is a tangent of *unit* length at A, and P is an arbitrary point on AB so that the length AP gives $t = \tan\theta$. The problem thus reduces to finding the length of the arc AP' corresponding to the length AP, which will give us θ as a function of t. *Yukti-bhasha* starts by dividing the line AB into a large number n of equal segments marking out $A_0 = A, A_1, A_2, \ldots, A_n = B$, with the line OA_i intersecting the arc at Q_i. The key insight is to realize that, as the half chord of a segment gets smaller and smaller, its length approaches the arc length to

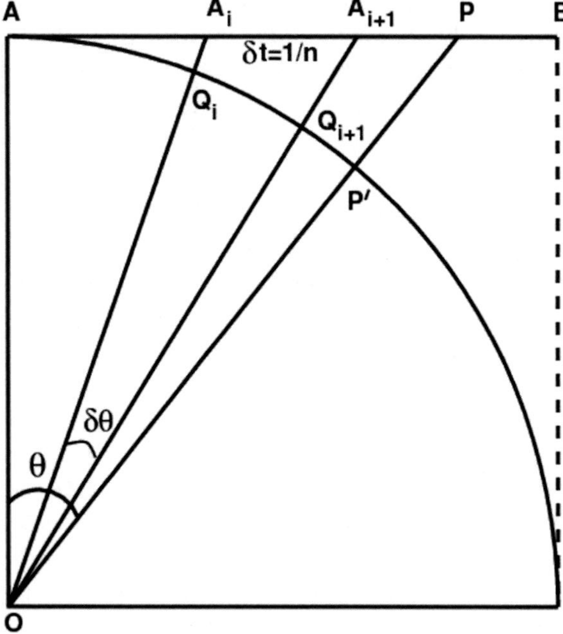

Fig. 20.1: Construction used in Yukti-bhasha to obtain the series expansion for π. See text for details.

arbitrary accuracy. In modern language, this is equivalent to the realization that $\sin(\theta/n) \to \theta/n$ as $n \to \infty$. In fact, there is an explicit statement in *Yukti-bhasha* to the effect that, "If the segments of the side [...] are very, very small, these half chords will be almost the same as arc segments." With this realization, and using some very innovative geometrical reasoning, one obtains the differential relationship $\delta\theta = [\delta t/(1+t^2)]$ between $\delta t = 1/n$ (given by $A_i A_{i+1}$) and $\delta\theta$ (given by $Q_i Q_{i+1}$), which is the relevant arc length in modern notation.

To obtain θ, one needs to sum the series $(1+i^2/n^2)^{-1}$ from $i = 1$ to n and finally take the limit $n \to \infty$. In *Yukti-bhasha*, this is done by expanding $(1+i^2/n^2)^{-1}$ in a series:

$$\delta\theta_i = \frac{1}{n} - \frac{i^2}{n^3} + \frac{i^4}{n^5} - \cdots .$$

(Interestingly, this is done by a powerful technique of recursive refining — called *samskaram* — in *Yukti-bhasha*.) This reduces the problem to one of finding the sums of powers of integers. While the sum of i^k for $k = 1, 2, 3$ was already known to Aryabhata, the results for higher values of k were not available. Once again Madhava cleverly uses the fact that he only needs the result for large n and *any subdominant term can be discarded.* This limiting process is carried out with extreme care in *Yukti-bhasha*, leading eventually to the result (called *samkalitam*, or 'discrete integration') which in modern notation gives

$$\lim_{n\to\infty} \frac{1}{n^{k+1}} \sum_{i=1}^{n} i^k = \frac{1}{k+1} .$$

This result — which is equivalent to integrating x^k and obtaining $x^{k+1}/(k+1)$ — was obtained using a procedure that reduces the sum of k th powers to the sum of the $(k-1)$th powers. This is a technique which we now call mathematical induction. From reading the relevant section of *Yukti-bhasha*, it is clear that its author was fully aware of the conceptual difference between this method of proof and the more traditional approaches usually based on geometrical reasoning; in fact, this is probably the first known instance of the use of mathematical induction in Indian mathematics. Once this step is completed, the result for the series can be obtained fairly easily. The clarity of the exposition regarding the ideas — like the limiting process and infinitesimals — shows how adept Madhava was at manipulating these to achieve his goal.

Equally innovative — or possibly even more so — was the approach used to obtain the series for sines and cosines. The first part of this analysis is again geometrical and closely parallels the discussion given above. By a similar reasoning,

Yukti-bhasha obtains the differential (again in modern notation) $\mathrm{d}(\sin\theta) = \cos\theta\,\mathrm{d}\theta$. In fact, this is done with a clever geometrical construction leading to a discretized difference formula for sine and cosine, expressed in the form:

$$\delta s_i \equiv s_{i+1} - s_{i-1} = 2s_1 c_i\ , \qquad \delta c_i \equiv c_{i+1} - c_{i-1} = -2s_1 s_i\ ,$$

where $s_i \equiv \sin(i\theta/2n)$ and $c_i \equiv \cos(i\theta/2n)$ for an angle θ divided into $2n$ equal parts. But, unlike in the case of the arctan series, one cannot now integrate this equation because the derivative of $\sin\theta$ itself is not a simple function to handle. The way *Yukti-bhasha* handles this situation is brilliant.

It converts the two coupled first order differential equations for sine and cosine into a second order difference equation by introducing the quantity (the 'second difference') $\delta^2 f_i \equiv \delta f_{i+1} - \delta f_{i-1}$, where f_i is s_i or c_i, and obtaining the result $\delta^2 f_i = -4s_1^2 f_i$. This is, of course, nothing but the differential equation $f'' + f = 0$ in discretized form, if we use our modern language. The next step is to solve this discretized equation by adding up the second differences to get first differences, and then adding up the first differences to get the function itself. This is again the discretized version of the fundamental theorem of calculus and again requires careful handling of limits. The relevant sections of *Yukti-bhasha* spend a significant time on this important issue, showing clearly that the author was aware of these subtleties.

This again leads to a *samkalitam* which, in the first instance, produces what we would now call an integral equation for $\sin\theta$ in the discretized form. This equation is solved by another finite *samkalitam* involving repeated substitution of the equation into itself, which requires calculating a set of sums of sums (*samkalita-samkalitam*, as you might have guessed the name should be!) defined recursively by $S_k(i) \equiv \sum_{j=1}^{i} S_{k-1}(j)$ with appropriate boundary conditions. The result for these sums of sums is given in *Yukti-bhasha*, but this is one instance in which the complete proof is not provided.

All this is done while keeping a finite n; in fact *Yukti-bhasha* generally uses the technique of delaying the step in which the limit $n \to \infty$ is taken for as long as possible. After the relevant sums are done, the time is ripe to let n tend to infinity and the k th term in the expansion of the sine series pops out with the right value $(-1)^k\theta^{2k+1}/(2k+1)!$ — one thus obtains the infinite series expansion for the sine. Translated into modern language, what has been achieved is equivalent to converting the differential equation for $\sin\theta$ into an integral equation of the form

$$\sin\theta = \theta - \int_0^\theta \mathrm{d}\phi \int_0^\phi \mathrm{d}\chi\ \sin\chi\ ,$$

and iterating it repeatedly to give

$$\sin\theta = \theta - \frac{\theta^3}{3!} + \cdots + (-1)^k \int_0^{\theta} \mathrm{d}\phi_1 \cdots \int_0^{\phi_{2k-1}} \mathrm{d}\phi_{2k} \sin\phi_{2k} \ .$$

During the entire process, *Yukti-bhasha* does not lose sight of the need to take the limit $n \to \infty$ right at the end. The word used in this context is *sunya-prayam* which translates to 'being similar to zero'; in today's language, we would have called it infinitesimal! The word *yathestam* is also used, meaning 'as one wishes', when talking about dividing lines into an *arbitrarily* large number of subdivisions, for example.

Box 20.2: Another Piece of Pi

Determining the circumference of a circle of given radius has intrigued all the ancient geometers and astronomers. Most of them used the procedure of inscribing and circumscribing the circle by polygons with many sides and approximating the circumference between the perimeters of the two polygons. Naturally, one must then find the side of the polygon as a function of the radius and the relevant results from the development of trigonometry.

While the approximate value $\pi \approx 22/7$ was available since antiquity, some better results were also known to many ancient mathematicians (see Box 4.1). Archimedes (287–212 BC) used a polygon with 91 sides to get the estimate $3\frac{10}{71} < \pi < 3\frac{10}{70}$. *Aryabhateeyam* (500 AD) states that the "circumference of a circle of diameter 20 000 is 62 832" leading to $\pi \approx 3.1416$, while Bhaskara ($\sim$ 1114–1185) takes the circumference to be 3927 for a circle of diameter 1250, obtaining $\pi \approx 3.14155$.

What is remarkable about *Yukti-bhasha* is that it does not try to give yet another *approximate* value for π, but an *exact* infinite series expansion — which is a giant conceptual leap forward. In fact, in addition to the result quoted in the text, *Yukti-bhasha* also gives an algebraic recursive method for finding π. In modern notation, this is based on using the recursion relation $x_{n+1} = x_n^{-1}\left[\sqrt{1+x_n^2} - 1\right]$ with $x_0 = 1$ and obtaining π as the limit

$$\pi = 4 \lim_{n\to\infty} 2^n x_n \ .$$

This method, however, requires the evaluation of square roots, which is not an easy procedure, while the series expansion given in the text, at the beginning of this chapter, contains no square roots.

Yukti-bhasha also works out the surface area and volume of a sphere by integration of the infinitesimal elements. For example, the surface area of a sphere of radius R is computed by considering a small strip located between the circles of latitudes θ and $\theta + \mathrm{d}\theta$ in, say, the upper hemisphere of the sphere. The area of this strip is first shown to be equal to $\mathrm{d}A = (2\pi R \sin\theta)(R\mathrm{d}\theta)$ by very careful and meticulous reasoning. In the discretized version, $\delta\theta$ becomes $\pi/2n$ and $\theta = i\pi/2n$, with $i = 0, 1, 2, \ldots, 2n$. The total area is obtained by summing over all the strips labeled by i and then taking the limit $n \to \infty$. This is, of course, the same as finding the integral of $\sin\theta$ to be $\cos\theta$. Though it was known that the first difference of the cosine is the sine, that information was not used, and this summation was done by effectively going through the original steps once again.

Historians are unsure whether the developments on the banks of the Nila influenced corresponding later developments in the West and — as far as we know today — there is no direct evidence that it did. (Kerala, of course, had a strong direct linkage with the West, for example through trading, from the very ancient days.) It is also somewhat unclear how the mathematical tradition of Nila faded after the seventeenth century, though several obvious historical factors can be identified as causes for its demise.

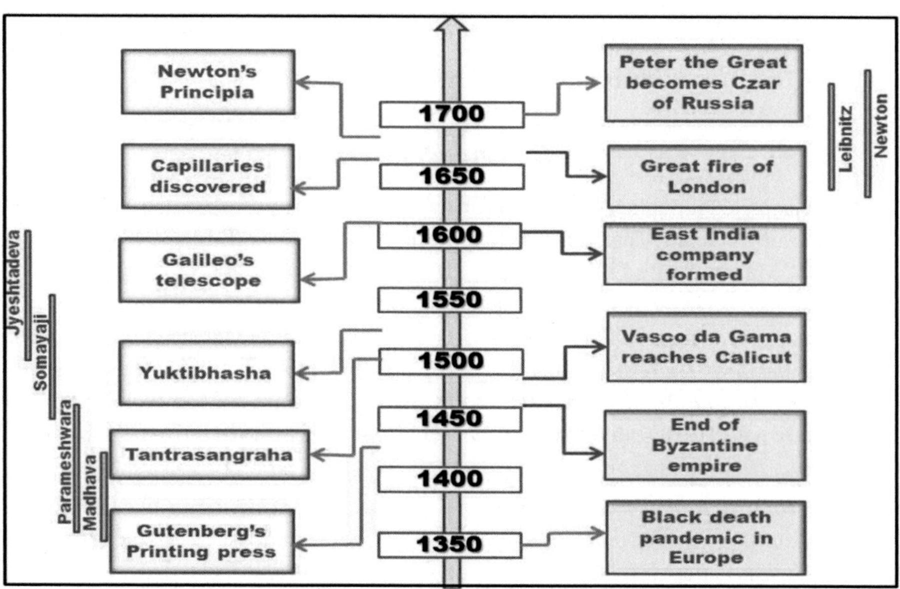

WHEN

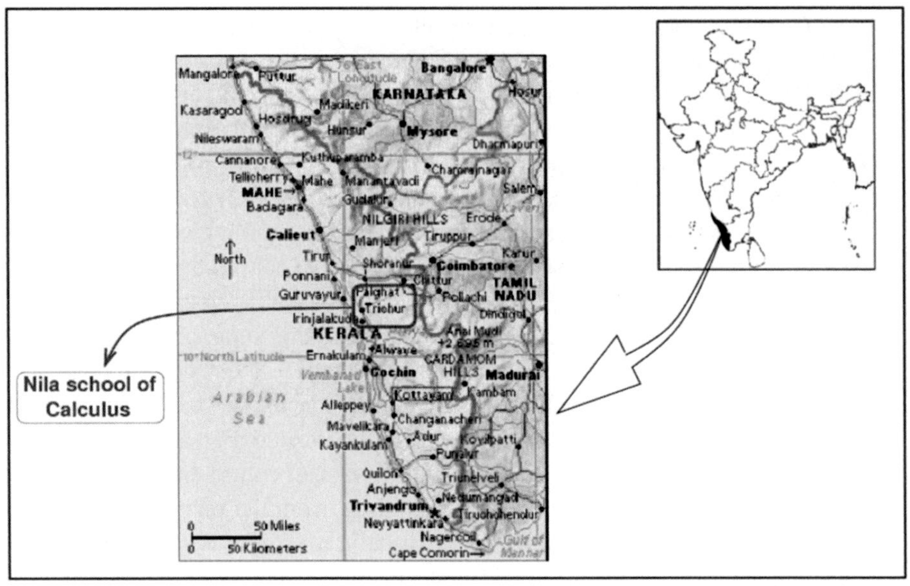

WHERE

Notes, References, and Credits

Notes and References

1. (a) The original publication was in Malayalam:
   Ramavarma (Maru) Thampuran and A. R. Akhileswara Aiyer (Eds.) (1948), *Yuktibhasa, Part I*, Magalodayam Ltd., Trichur, Kerala.
   (b) For the English version with detailed commentary, see:
   Sarma, K.V. et al., (2008), *Ganita-Yukti-Bhasa (Rationales in Mathematical Astronomy) of Jyesthadeva. Sources and Studies in the History of Mathematics and Physical Sciences (in English and Malayalam), Part I (Mathematics), Part II (Astronomy)*, Springer (Jointly with Hindustan Book Agency, New Delhi, India) [ISBN 978-1-84882-072-2].
2. Whish, Charles (1834), *Hindu Quadrature of the circle and the infinite series of the proportion of the circumference to the diameter exhibited in the four Sastras, the Tantra Sahgraham*, Transactions of the Royal Asiatic Society of Great Britain and Ireland, Royal Asiatic Society of Great Britain and Ireland **3** (3), 509–523, [doi:10.1017/S0950473700001221].
   This paper has been reproduced as an Appendix in:

Bhanu Murthy, I.S. (1992), *A modern introduction to ancient Indian mathematics*, New Age International Publishers, New Delhi [ISBN 81-224-0371-9].

3. For more details, see, e.g.,

Divakaran, P. P. (2007), *The First textbook of Calculus: Yuktibhasha*, J. Indian Philosophy **35**, 417.

Divakaran, P. P. (2018), *The Mathematics of India: Concepts, Methods, Connections* Springer-Hindustan Book Agency, [ISBN-10: 9811317739].

Rajeev, S. G. (2005), *Neither Newton nor Leibnitz: The pre-history of calculus in medieval Kerala*, Lectures at Canissius College, Buffalo, New York [available at http://www.pas.rochester.edu/~rajeev/papers/canisiustalks.pdf].

Ramasubramanian, K. and Srinivas, M. D. (2010), *Development of Calculus in India*, in: Studies in the History of Indian mathematics, pp. 201–286, Cult. Hist. Math. 5, Hindustan Book Agency, New Delhi; available at http://www.ms.uky.edu/sohum/ma330/files/india_calculus.pdf

Plofker, Kim (2009), *Mathematics in India,* Princeton University Press, Princeton, USA [ISBN 978-0691120676].

George Gheverghese Joseph (2016), *Mathematics: Engaging with the World from Ancient to Modern Times*, World Scientific, Singapore [ISBN 9781786340603].

21

Story of the Calendar

Accurate time-keeping at all levels has played a crucial role in the evolution of science. An important off-shoot of this which directly affects the common man is the development of calendar systems [1]. While this may not be considered a development *within* science from the narrow, conventional view, it certainly had a symbiotic relationship with it — and it is *inextricably connected with the knowledge of astronomical phenomena* — thereby deserving its own discussion here.

Most of us take a good calendar for granted, without giving it a second thought. If you want to understand how non-trivial the development of a calendar system is, you have to think about the primordial origins of our calendar. Suppose you were living in Egypt 3000 years ago and your friendly king had appointed you as his official priest. He would have liked to know several things from you: When should the farmers sow the seeds? When should he move to that nice palace on top of the mountains? When should he offer his daily prayers? Since you are enjoying the benefits of being God's representative in the country, you had better predict things right. Any repeated disasters in your predictions – and you might end up as breakfast for the royal crocodiles.

Let us see how you can go about developing a system of time-keeping which will keep you in the king's good books. Certain obvious regularities in nature come to your aid. The first is the concept of the day: you see light and darkness at regular intervals, connected with the rising and setting of Ra — or, in more familiar terminology, the Sun. Next is the regularity of the changes in the shape of a second celestial object, the Moon: it starts from nothing, reaches full splendour,

T. Padmanabhan and V. Padmanabhan, *The Dawn of Science*,
https://doi.org/10.1007/978-3-030-17509-2_21

and vanishes back to nothing. Third is the more subtle, but significant, regularity of the seasons. From experience you know that if winter comes, spring can't be far behind. All you have to do is develop a system which will enable you to predict these changes accurately in advance.

The trouble begins exactly here. It makes sense to keep the day — say, the time between two consecutive sunrises as the basic unit. Now you need to know how many days it takes for the Moon to finish one cycle of phases. To your utter dismay this turns out to be something *between* 29 and 30 days — pretty close to $29\frac{1}{2}$ days. That is quite inconvenient if you want to use the Moon's cycle as the second unit of time. (Let us call it a 'month,' for convenience.) To keep in phase, a month should have $29\frac{1}{2}$ days. If you declare a month to be 29 or 30 days, it becomes useless pretty soon — in the sense that the beginning of months and the beginning of the Moon's cycle will quickly fall out of step!

The troubles continue at the next level as well, when you try to connect the lunar cycle with seasonal changes. The time between two consecutive seasons — or between the floods in your nearby Nile — turns out to be between 12 and 13 months irrespective of whether your months have 29 days or 30 days. Suppose you had declared your month to be of 29 days and the year to be 12 months. Then you find that the seasons come late by nearly 17 days each year. In four years you are off by nearly 2 months, and if the farmers are sowing their fields based on months, even your gods won't be able to protect you from the king's wrath! Making a calendar has become an occupational hazard.

The dilemma described above was faced by the priests and the time-keepers of every ancient civilization. Each of them came up with a calendar system best suited for their needs, mostly religious and agricultural, and sometimes social. The story of the calendar is the story of man's attempt to develop some order in time-keeping.

Two of the major calendar makers among the ancient civilizations in the West were the Egyptians and the Babylonians. (There is some evidence that the Sumerians had also worked out a calendar system, but we do not have a similar amount of detail about it.)

The Egyptian priests had a simple and elegant calendar. They realized that it was too confusing to worry about *both* the Sun and the Moon. After all, seasons are important for an agricultural society, but the phases of the Moon seldom seemed to affect anything. So they stuck with the Sun and their year consisted of 12 months (each of 30 days) and an additional 5 days added at the end and treated as holidays. This kept the calendar broadly in tune with the seasons. The new Moon could fall on any day of a month and the Egyptians didn't care. The most important seasonal event in ancient Egypt was the flooding of the Nile, which watered the fields and fertilized them with silt, keeping the "Gift of the Nile" prosperous. This flooding takes place in mid-July (what we call July *now*, that is) and it coincides with the appearance of Sirius — the brightest star, called Sothis by the Egyptians — in the east just before dawn. That certainly looked very auspicious, and the Egyptians kept it as the beginning of the year.

The 'real' year, of course, is not 365 days long. It lasts 365.242 199 days or rather 365 days, 5 hours, 48 minutes, and 46 seconds. To a first approximation, we say that it is 365 and 1/4 days long. This means that each year, the Egyptian year of 365 days falls 1/4 of a day behind the Sun. In 4 years it is 1 day behind schedule, and so on, until in 1460 years it has gone through one complete circuit. Thus, 1460 astronomical ('solar') years will be 1461 Egyptian years. The Egyptians knew about this and called the 1461 years the Sothic cycle. In fact, Ptolemy III Euergetes (284–222 BC) — the Egyptian monarch of Greek origin — suggested adding an extra day every fourth year! The traditional priesthood vehemently opposed it and Ptolemy seems to have given up trying to make them see sense.

The Babylonians, however, were more demanding. They wanted to keep the correspondence with the Moon as well. As you probably realized before — in your short tenure as the court priest — this is quite impossible with 29-day or 30-day months. You could do slightly better with a 29.5 day month, which is what the Babylonians did. They had the year made of 6 months of 29 days and 6 of 30 days in alternation. That doesn't however, solve the problem completely. You can see that this year totals to 354 days while the seasons follow the 365 day cycle. One could once again have just added the 11 days at the end as holidays, but that was not acceptable to the Babylonians, because that would make the months go out of step with Moon's phase. Even as it was, there was some trouble in store. One lunar month is actually 29.5306 days rather than 29.5 days. In other words, a true lunar year is $12 \times 29.5306 = 354.37$ days rather than 354 days. It becomes a big deal when these errors accumulate. Suppose then that you start your year on a new Moon day and have a year of alternating 29 and 30-day months. In 3 years time, the year will start a day before the new Moon, and in 6 years, 2 days before, etc. This can seriously offend religious sensibilities.

Faced with these troubles, the Babylonians came up with a complicated correction scheme. They noticed that 19 solar (seasonal) years contain just about 235 lunar months. The 19 solar years will have $19 \times 365.2422 = 6939.60$ days while 235 lunar months will have $235 \times 29.5306 = 6939.69$ days, so the two are pretty close. So, if we start a solar year with a new Moon and wait, the 20th solar year will again start on the new Moon. Now, these 235 lunar months are the same as 19 Babylonian lunar years plus 7 lunar months ($235 = 19 \times 12 + 7$); or equivalently,

19 *lunar* years added to 7 *lunar* months is the same as 19 seasonal *solar* years. So the simplest correction needed to make lunar years agree with seasons would be the following: Let the lunar year go on for 19 years. Now, we will be 7 months behind in seasons. Just add these 7 months to the 19 years (i.e., making the 19th year 19 months long) and you are back with the seasons as though nothing happened.

This is alright except that seasons which are even a few months behind schedule can cause havoc for the common man. So, the Babylonians distributed these 7 extra months within the 19-year cycle as evenly as possible. They added one month to the 3rd, 6th, 9th, 11th, 14th, 17th, and 19th year of the cycle, making it have 12 twelve-month years and 7 thirteen-month years. It is quite intricate but the year was never more than 20 days off with respect to the Sun. The 19-year cycle is called the Metonic cycle after the Greek astronomer, Meton, who calculated it around 430 BC. The present day Jewish calendar is a luni-solar calendar developed from the Babylonian calendar described above.

Though the Metonic cycle came from a Greek astronomer, the Greeks themselves were pretty terrible calendar makers. They had a lunar year of 354 days and used to add corrections whenever they found the seasons to misbehave. Besides, each city state had its own scheme, adding to the confusion. By the years 87 BC to 84 BC, the Roman conquest of Greece was complete. This, however, didn't help the calendar because the Roman priests were as terrible as the Greek ones in correcting the calendar. This was to be expected because the Roman calendar was quite unsystematic.

Box 21.1: Other Calendar Systems

Several other ancient civilizations had their own calendars. Notable among them are the Chinese, Mayan, and Aztecs.

Inscriptions show that the Shang dynasty in China (fourteenth century BC) already knew that one year is $365\frac{1}{4}$ days and that the time between two new Moons is $29\frac{1}{2}$ days. They also knew about the Metonic cycle at least a century before Meton. Their calendar reflected these facts: months were of 29 and 30 days with extra intercalation of months whenever necessary. Later on, around the third century BC, they switched over to a system involving '24 points' which are 15° apart on the ecliptic. It takes about 15.2 days for the Sun to travel from one point to another. Thus two 'points' share 30.4 days, which is a little longer than the lunar month of 29.5 days. The Chinese wanted to retain the lunar structure of the months as well and achieved this by careful intercalations.

The Mayan system was much more complicated. They had a ritual cycle of 260 named days (called tzolkin) and a year of 365 days! Together, they form a longer cycle of 18980 days, at the end of which the pattern repeats. The 260-day cycle is formed by meshing the numbers 1 to 13 with an ordered series of 20 names, whilethe 365-day cycle consists of 18 named months, each of 20 days, with an additional 5 days of evil omen called Vazeh. It is clear that these cycles could not be in phase with any astronomical phenomenon; also the Mayans had a very complicated way of recording dates.

The Aztecs had almost the same structure for their calendar. The names of the 20 months (in the 260-day cycle) also bear a striking resemblance to those of the Mayans. However, they used a more primitive system for naming the years and days. As a result, the record of events kept by the Aztecs is more ambiguous.

Originally, Romulus — the founder of Rome — (700 BC?) had a year of 10 months: March, May, Quintilus (July), and October with 31 days each, and April, June, Sextilies (August), September, November, and December with 30 days, totaling to a 304-day year. Incidentally, this explains the nomenclature of ‘Sept’ (seven), ‘Oct’ (eight), etc. Naturally, the 304 days had no astronomical significance and the early Romans arbitrarily added ‘winter gaps’ to keep pace with the seasons.

Emperor Pompilius (753–673 BC) produced some sense of order by adding 2 months, January and February, and redistributing the days as follows: he gave January 29 days, February 28 days, and took one day off from all the 30-day months, making them all 29 days. This produced a calendar with four 31-day months, seven 29-day months and one month with 28 days. The year was 355 days long and the seasonal corrections were supposed to have been made by periodically adding an extra 22 or 23 days in February. In practice, however, politicians and priests would add or subtract days to suit their needs, creating a fair amount of chaos.

This pathetic situation continued until the time of Julius Caesar. He noticed the remarkable simplicity of the Egyptian calendar during his Egyptian campaigns, and decided to put an end to the local Roman nonsense. He imported the Egyptian astronomer, Sosigenes, to help him in this matter. In the year 46 BC, they decided to add 90 extra days to that year — 23 in February and 67 between November and December — making it 445 days long. This brought the year in phase with the seasons and ended the previous confusion (though, ironically enough, 46 BC was called the 'year of confusion'). In order to keep the year in phase with the seasons in the future, Caesar also ordered that February should have an extra day every fourth year, which would be a leap year. (The 365-day year is 52 weeks and 1 day long. So if 10th March is a Sunday in one year, it will be a Monday the next year and so on. If a leap year intervenes, it will 'leap' and will fall on a Tuesday in the next year; hence the name.)

Incidentally, the Romans did not number the days of a month sequentially as $1, 2, \ldots$. Instead, they counted back from the three fixed dates of reference in each month, called the Nones (5th or 7th), the Ides (13th or 15th) — which led to the famous Shakespeare line, "Beware the Ides of March" — and the Kalends (1st) of the next month. The Nones of April was the 5th, and the Ides the 13th (see Fig. 21.1). The last day of April was the *pridie Kalendas Maias*, i.e., "the day before the Kalends of May". Roman counting was inclusive; April 9 was *ante diem V Idus Aprilis*, "the 5th day before the Ides of April," usually abbreviated a.d. V Id. Apr; April 23 was IX Kal. Mai., "the 9th day before the Kalends of May," on the Julian calendar (VIII Kal. Mai. on the pre-Julian calendar).

The noble intentions of Caesar were almost put in jeopardy by his Pontifices who were accustomed to 'inclusive numbering.' That is, they merrily added one day every *third* year rather than every fourth. This continued unnoticed for 36 years, during which time 12 days had been added rather than 9. Fortunately, emperor Augustus corrected this by omitting the extra days between 8 BC and 4 AD. He also clarified the leap year prescription, bringing the Julian calendar properly into operation.

Caesar and Augustus also changed the days of the months. Caesar added two days to January, September, and November, and one-day each to February, April, June, Sextilis (August), and December. He reduced October by one day. This produced a year with months having alternately 30 and 31 days (March 31, April 30, etc.) except for February, which had 29 (or 30 days in a leap year). This totals 365 days.

Human ego didn't let such an orderly state of affairs continue. The Roman senate altered the names Quintilis to Julius ('July') in honour of Caesar and Augustus persuaded them to change Sextilis to Augustus in his honour. Since he wanted August to be at least as long as July, he increased August to 31 days, reducing February to 28 days. This, however, produced 3 months in succession — July, August, and September — with 31 days. To avoid this he reduced September and November to 30 and increased October and December to 31. That is how we got our crazy system of "30 days hath September . . . ."

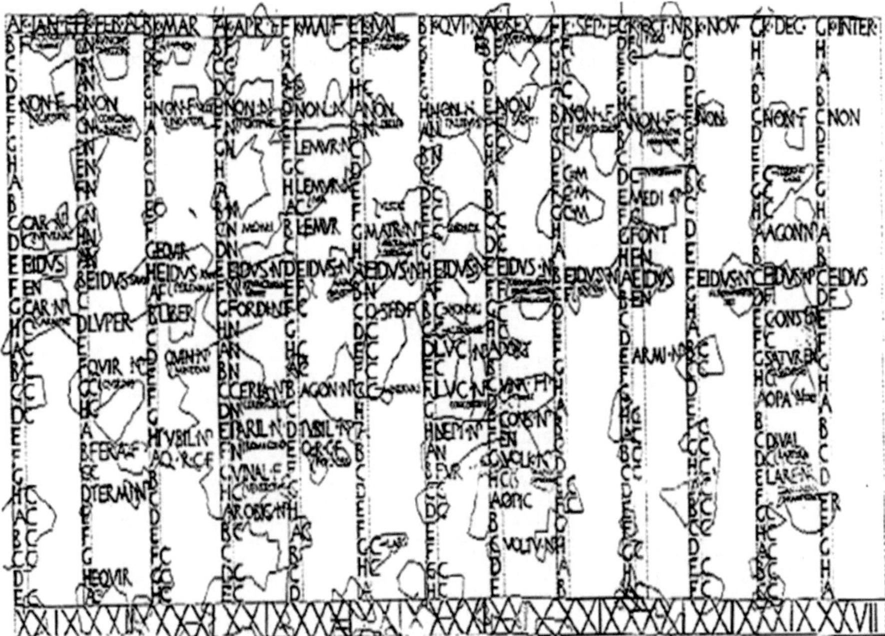

Fig. 21.1: A black and white reproduction of a fragmentary fresco [4] of a pre-Julian Roman calendar (the original fresco has red letters on a white background, now kept in the Palazzo Massimo) found in the ruins of Nero's villa at Antium (circa 60 BC) with the seventh and eighth months still named Quintilis ('QVI') and Sextilis ('SEX') and an intercalary month ('INTER') in the far right hand column. The lengths of January to December are 29, 28, 31, 29, 31, 29, 31, 29, 29, 31, 29, 29 days each in years without the leap month. The letter F on certain days signifies that they were 'dies fasti', viz., days on which legal action was allowed; the letter N is for 'dies nefasti' or days on which legal action was not allowed; the letter C is for days on which citizens could vote on different matters; the letters EN were considered a mixture between C and F days, with the mornings used for one purpose and the afternoons for another. The letters NP denote public holidays, while QRCF are days when the 'king' (rex sacrorum) could convene an assembly.

Nature, however, did not leave the calendar makers in peace. The trouble, of course, is that the seasonal year is not exactly 365.25 days long. It actually lasts 365.242 199 days. So the Julian year is about 11 minutes and 14 seconds too long. A tiny amount, but it means that the Julian calendar gradually crawls forward to gain one full day in 128 years, and this ended up causing more trouble.

CALENDARIVM GREGORIANVM PERPETVVM.

Orbi Chriſtiano vniuerſo à GREGORIO XIII. P. M. propoſitum. Anno M. D. LXXXII.

GREGORIVS EPISCOPVS SERVVS SERVORVM DEI AD PERPETVAM REI MEMORIAM.

INTER grauiſſimas Paſtoralis officij noſtri curas, ea poſtrema non eſt, vt quæ à ſacro Tridentino Concilio Sedi Apoſtolicæ reſeruata ſunt, illa ad finem optatum, Deo adiutore perducantur. Sane eiuſdem Concilij Patres, cum ad reliquam cogitationem Breuiarij quoque curam adiungerent, tempore tamen excluſi rem totam ex ipſius Concilij decreto ad auctoritatem & iudicium Romani Pontificis retulerunt. Duo autem Breuiario præcipue continentur; quorum vnum preces, laudesque diuinas feſtis, profeſtisque diebus perſoluendas complectitur, alterum pertinet ad annuos Paſchæ, feſtorumque ex eo pendentium recurſus, Solis, & Lunæ motu metiendos: Atque illud quidem felicis recordationis Pius V. prædeceſſor noſter abſoluendum curauit, atque edidit. Hoc vero, quod nimirum exigit legitimam Calendarij reſtitutionem, iamdiu à Romanis Pontificibus prædeceſſoribus noſtris, & ſæpius tentatum eſt, verum abſolui, & ad exitum perduci ad hoc vſque tempus non potuit, quod rationes emendandi Calendarij, quæ à cæleſtium motuum peritis proponebantur, propter magnas, & fere inextricabiles difficultates, quas huiuſmodi emendatio ſemper habuit, neque perennes erant, neque antiquos Eccleſiaſticos ritus incolumes (quod in primis hac in re curandum erat) ſeruabant. Dum itaque nos quoque credita nobis, licet indignis, à Deo diſpenſatione freti, in hac cogitatione, curaque verſaremur, allatus eſt nobis liber à dilecto filio Antonio Lilio artium, & medicinæ doctore, quem quondam Aloyſius eius germanus frater conſcripſerat, in quo per nouum quendam Epactarum Cyclum ab eo excogitatum, & ad certam ipſius aurei numeri normam directum, atque ad quamcunque anni ſolaris magnitudinem accommodatum, omnia, quæ in Calendario collapſa ſunt, conſtanti ratione, & ſeculis omnibus duratura, ſic reſtitui poſſe oſtendit, vt Calendarium ipſum nulli vnquam mutationi in poſterum expoſitum eſſe videatur. Nouam hanc reſtituendi Calendarij rationem exiguo volumine comprehenſam ad Chriſtianos Principes, celebrioresque vniuerſitates paucos ante annos miſimus, vt res, quæ omnium communis eſt, communi etiam omnium conſilio perficeretur; illi cum, quæ maxime optabamus, concordes reſpondiſſent, eorum nos omnium conſenſione adducti, viros ad Calendarij emendationem adhibuimus in alma Vrbe harum rerum peritiſſimos, quos longe ante ex primarijs Chriſtiani orbis nationibus delegeramus: Ii cum multum temporis, & diligentiæ ad eam lucubrationem adhibuiſſent, & Cyclos tam veterum, quam recentiorum vndique conquiſitos, ac diligentiſſime perpenſos inter ſe contuliſſent, ſuo, & doctorum hominum, qui de ea re ſcripſerunt, iudicio hunc præ cæteris elegerunt Epactarum Cyclum, cui nonnulla etiam adiecerunt, quæ ex accurata circumſpectione viſa ſunt ad Calendarij perfectionem maxime pertinere.

Fig. 21.2: The first page [5] of the papal bull "Inter Gravissimas" [2] by which Pope Gregory XIII introduced his calendar reform. The bull refers to "the explanation of our calendar" and to a canon related to a dominical letter. There were six chapters of explanatory rules ('canons') which accompanied the bull. Some of these (canons 1, 2, 4) refer to a book entitled *Liber novae rationis restituendi calendarii Romani* for a fuller explanation of the tables than was contained in the canons and the bull. But it appears that the *Liber novae* . . . never saw the light of day [3]!

Many Christians — especially the Romans who embraced Christianity — were using the Julian calendar to calculate the date of Easter. The rule used to calculate Easter was quite complicated, but depended crucially on the vernal equinox falling on 21st March. This formula was worked out at the Christian Council of Nicaea in AD 325. But alas, by AD 1263, the Julian year had gained 8 days and the vernal equinox was on 13th March. It was clear that if this state of affairs continued, Easter, tied to the vernal equinox, would come to be celebrated in mid-summer and Christmas in spring. In 1263, Roger Bacon wrote to Pope Urban IV, explaining the situation.

The church took nearly three centuries to do something about it! The Papal 'Bull' (Fig. 21.2) was finally issued on 15 February 1582 by Pope Gregory XIII. (He was educated on calendar-making by the Jesuit astronomer Christoper Clavius and Aloysius Lilius). Gregory summarily dropped 10 days changing 5 October 1582 to 15 October 1582. This brought the calendar in phase with the Sun and the vernal equinox of 1583 did then fall on 21 March.

It was, of course, necessary to prevent this from happening again. Since the Julian year gains one day in 128 years, it gains 3 days in 384 years, or roughly, every 4 centuries. So if we omit 3 leap days in every 400 years, all will be well. Gregory devised a simple rule for this. Consider the century years 1500, 1600, 1700, etc. In the Julian system, all century years (which are always divisible by 4) would have been leap years. Since there are 4 century years in every 400 years, we can keep 3 of them ordinary years and only one as a leap year. This, Gregory reasoned, would ensure automatic subtraction of 3 days from the Julian calendar every 400 years. This is the rule we now follow: any *non-century* year is a leap year if it is divisible by 4. A century year is a leap year only if it is divisible by 400.

Box 21.2: The October Revolution is Celebrated in November!

By the time the papal bull on the calendar reform was issued, Europe was divided into Catholic and Protestant nations. Each protestant country had to decide whether to follow the Pope and be right or defy the papal bull and stick with the literally out-dated Julian calendar. Probably not surprisingly, many Protestant nations preferred to stay out of step with the Sun than to accept a correction by the Pope!

It took time for them to give in, and by 1700, Denmark, Holland, and Protestant Germany adopted the Gregorian calendar. Britain and the American colonies held out until 1752! By then, they had to drop 11 days which they did by changing the date 2 September 1752 to 13 September 1752. Incidentally, this created public confusion and displeasure, especially because landlords charged a full quarter rent even though September was 11 days short. (Some history books say that people rioted after the calendar change, asking that their "eleven days" be returned. However this is very likely a myth, based on sources like the painting by William Hogarth, shown in Fig. 21.3.)

This late reform had, for example, shifted George Washington's birthday. He was born on 2 February 1732 (Gregorian calendar) but the date recorded by the family was the Julian date 11 February 1732. When the changeover took place, Washington quietly changed his birthday.

Matters were worse in the Soviet Union. The Russian Orthodox Church clung on to the Julian calendar — with the accumulated errors, of course — until as late as 1918. After the Bolshevik revolution, they dropped 13 days to bring their calendar in line with the rest of Europe. That is why the "October revolution" is now celebrated on 7 November.

Every 400 years, the Julian calendar has 100 leap years and a total of 146 100 days. In the same 400 years, the Gregorian calendar has 97 leap years and hence a total of 146 097 days; 400 seasonal solar years has 146 096.88 days. Thus, the Gregorian year gains only 0.12 day, or about 3 hours in 400 years, while the Julian calendar would have gained 3.12 days. A little calculation shows that the Gregorian calendar will gain one full day in about 3400 years. So, around AD 5000, we may have to 'drop' one day to be in phase with the seasons.

Unfortunately, the reforms came a little too late historically. By 1582, a large fraction of Europe had turned Protestant and there followed the usual clash between a sensible reform and strong religious sentiments. Different countries switched to the new calendar at different times, leading to rather amusing results (see Box 21.2).

Fig. 21.3: This painting [6] by William Hogarth (1755), part of a series known as "The Humours of an Election", contains the famous "Give us our Eleven Days" protest slogan against the Gregorian calendar. This slogan is at the lower right of the painting, on a black banner on the floor under a person's foot! This is most probably based on the 1754 Oxfordshire elections, in which the 1752 calendar change was one of a number of issues brought up by Tory opponents to the Whig candidate, who had been influential in passing the calendar law. The painting by William Hogarth is rich in details: for example, it has Tory and Whig agents, both attempting to bribe an innkeeper to vote for them; and in a reference to the anti-semitism of the crowd behind, a Jewish peddler is shown as being employed by another agent who is offering jewels and ribbons to the wives of voters.

Box 21.3: The Ubiquitous Friday the Thirteenth

As explained in the text, century years are leap years only if they are divisible by 400 (i.e., 2100 is not a leap year but 2000 is). Because of this rule, the entire day-pattern of the Gregorian calendar repeats in a cycle of 400 years. (It is easy to see that 400 years contain an integral number of weeks.) Thus, by examining a convenient span of 400 years, we can decide about any day–date pattern.

Consider, for example, the superstition that the combination 'Friday the thirteenth' is especially sinister. In a 400-year cycle, the 13th of a month falls on Sunday 687 times, on Monday and Tuesday 685 times each, on Wednesday 687 times, on Thursday and Saturday 684 days, but on Friday 688 times! So we will have the 13th of months falling more frequently on Friday than on other days!

Most of the commercial, political, and social events of the world follow the Gregorian calendar today. Religious groups decide on their festivals by their own calendars. It is probably worth mentioning two of them — the Hindu and the Muslim calendars.

The ancient Vedic text, Taittireya Samhita, divides an approximate solar year of 360 days into 12 lunar months of 27 or 28 days. All resulting discrepancies are resolved by introducing periodic corrections, and in particular, one leap month every 60 months. The ecliptic circle in the sky, through which the Sun moves, is divided into 12 parts, each corresponding to a sign of the zodiac and spanning 30 degrees in the sky. Each of these 12 zodiacs are further divided into $2\frac{1}{4}$ 'Nakshatras' (Sanskrit for 'stars'), thereby giving a division of the ecliptic circle into 27 Nakshatras as well. The 12 months are decided and named on the basis of the position occupied by the Sun in the zodiac. A day can be assigned a Nakshatra based on the position of the Moon among the 27 Nakshatras. However, in the ancient Hindu tradition, the basic unit is not so much a day of the week but a 'Nakshatra' or 'Thithi.' Roughly speaking, a 'thithi' is one thirtieth of a lunar month and repeats roughly twice each month, once between the new Moon and full Moon and once between the full Moon and the new Moon. These two halves are distinguished by two 'Pakshas' — the Krishna Paksha (waning phase) and the Shukla Paksha (the waxing phase). Thus ancient dates are invariably given by specifying the month ('masa'), the fortnight ('the paksha') and the Thithi. The system requires computation of the Almanac every year, keeping the astronomical facts in mind.

Box 21.4: How Jesus Was Born a few Years Before Christ

In India, the Vikrama Era and the Saka Era started in 58 BC and 78 AD, respectively. The Islamic calendar starts from 622 AD, the year Mohammad migrated from Mecca to Medina. But how did the Western Eras start?

The ancient Greeks made use of the Olympian games — celebrated every four years — for this purpose. Years were marked as the 'third year after the tenth Olympics', etc. This unnecessarily complicated scheme was modified by Seleucus, who kicked off the 'Seleucid Era' after his victory at the battle of Gaza (1st year of the 117th Olympics). This was later adopted by several Jewish communities.

In parallel was the Roman Era, beginning from the 4th year of the 6th Olympics, in which the city of Rome was founded. This year is denoted by A.U.C. (Anno Urabis Conditae' or 'year of founding the city'). Early Christians, anxious to show that the biblical evidence antedates Rome, claimed that Abraham was born 1263 years before the founding of Rome. This was taken to be the first year in the Era of Abraham. Medieval Jews went further and even calculated the 'year of creation' or 'year of earth' (the Mundane Era); then 1989 AD becomes Jewish year 5750.

In about 1288 A.U.C., a Syrian monk named Dionysius, using biblical data, calculated that Jesus must have been born in 754 A.U.C. That was adopted as the starting point for Anno Domini ('the year of the Lord') in the time of Charlemagne. Unfortunately, Dionysius was off the mark by about 4 years. Jesus was supposed to have been born during the days of Herod who died in 750 A.U.C. So Jesus was born at least by 4 BC — that is, four years Before Christ.

The Islamic calendar, on the other hand, is completely lunar. The year consists of 12 months alternating between 29 and 30 days. This is identical to the ancient Babylonian calendar, except that Babylonians had additional months in a 19 year cycle, keeping the calendar in phase with the seasons. When Caliph Omar — follower of Mohammad — took over this luni-solar calendar from the Middle East, he dropped the idea of extra months. Thus, every Islamic year has only 354 days (sometimes one extra day is added to the last month), so the months do not keep to the seasons, except that 33 Islamic years average to about 32 Gregorian years. The

Muharram festival — which is the Islamic new year's day — shifts through the Gregorian calendar. The ninth month, Ramadan, is especially holy for Muslims and is observed as a month of fasting. According to the Quran, Muslims must see the new Moon crescent before beginning the fast. This is important because — as we explained earlier — the months of a lunar year do not automatically begin on the new Moon day. Most Islamic countries use the Gregorian calendar for civil and practical affairs, but use the Islamic calendar to decide on the dates of religious festivals.

Notes, References, and Credits

Notes and References

1. For more on calendar systems, see, e.g.,
   Steel, Duncan (1999), *Marking Time: The Epic Quest to Invent the Perfect Calendar*, Wiley and Sons, USA [ISBN 978-0471298274].
   Duncan, David Ewing (1999), *Calendar: Humanity's Epic Struggle to Determine a True and Accurate Year*, Harper Perennial, New York [ISBN 978-0380793242].
   Blackburn, Bonnie and Holford-Strevens, Leofranc (1999), *The Oxford Companion to the Year: An Exploration of Calendar Customs and Time-Reckoning*, Oxford University Press, Oxford [ISBN 978-0192142313].
2. An English translation of the Latin 'Inter Gravissimas' can be found here:
   http://myweb.ecu.edu/mccartyr/intGrvEng.html
3. See
   Nothaft, Philipp E. (2018), *Scandalous Error: Calendar Reform and Calendrical Astronomy in Medieval Europe*, Oxford University Press, Oxford, UK [ISBN 978-0198799559].

Figure Credits

4. Figure 21.1 courtesy: Wikimedia Commons/Public domain,
   https://en.wikipedia.org/wiki/File:Roman-calendar.png (from public domain).
5. Figure 21.2 courtesy: Wikimedia Commons/Public domain,
   https://en.wikipedia.org/wiki/File:Inter-grav.jpg (from public domain).
6. Figure 21.3 courtesy: William Howarth, Wikimedia Commons/Public domain,
   https://en.wikipedia.org/wiki/File:William_Hogarth_028.jpg (from public domain).

22

And Then All Was Light

In the entire history of physics, there have been four men — Archimedes, Galileo, Newton, and Einstein — who belonged to a class of their own, far above the rest. Of the four, Isaac Newton (1643–1721) belonged to a period neither too far in the past nor too close to the present, which has led to a trail of fables and (usually exaggerated) anecdotes about him. In this chapter, we will explore some aspects of the life and work of Newton.

Newton was born on 4 January 1643, and was not really a Christmas baby as is popularly believed. This date is based on the modern Gregorian calendar, which is not only in conformity with Nature but was also being followed all over Europe (except England) at the time of Newton's birth. England, however, was sticking with the outdated Julian calendar (see Chap. 21), according to which Newton was born on 25 December 1642, thereby giving him the glamour of being a Christmas baby [1]. His father was illiterate (though quite prosperous) and had died three months before Newton was born. His mother married again when he was three and Newton was left with his grandmother for about nine years. Much later, in 1662, Newton recorded that he had actually threatened his mother and stepfather "to burne them and the house over them". This statement has provided enough fodder for psychoanalysts to attribute every trait of Newton to the lack of motherly love in his childhood!

His mother returned to him after nine years, when her second husband had died, and brought with her the three children from the second marriage. Some attempt was then made to persuade Newton to manage the family estate, which would

T. Padmanabhan and V. Padmanabhan, *The Dawn of Science*,
https://doi.org/10.1007/978-3-030-17509-2_22

Fig. 22.1: Portrait [9] of Isaac Newton (1643–1721), one of the four giants in physics who form an elite set — the other three being Archimedes, Galileo, and Einstein.

have been disastrous — both for the estate and for Newton. Fortunately, Newton's inabilities in these matters were detected quite early and he was instead sent to the Free Grammar School of Grantham to prepare for entering the University of Cambridge [2].

In June 1661, Newton entered Trinity College, Cambridge, which remained his home for most of the next 35 years. The official curriculum in Cambridge in those days was distinctly Aristotelian, resting on the geocentric view of the universe and dealing with Nature in very qualitative terms. However, the scientific revolution was fairly advanced in Europe and the writings of Copernicus, Kepler, Galileo, and Descartes were available to those who were interested in them. By 1664, Newton had read and thought deeply about many of these works. In particular, he was strongly influenced by the mechanistic philosophy advocated by Descartes. In one of his notebooks, Newton had entered the slogan *Amicus Plato, amicus Aristotle, magis amica veritas* (Plato is my friend, Aristotle is my friend, but my best friend is truth). Clearly, an original thinker had been born, in spite of the best efforts of his university.

Around this time, Newton had mastered most of classical mathematics and was moving into new territories. The basic ideas of the binomial theorem and calculus were probably already in place, though the world came to know of these only much later. He received his bachelor's degree (which was rather a formality, merely recording his completion of four years at Trinity) in 1665, one year after he was elected a Scholar of Trinity. But his stay in Cambridge came to an abrupt end in 1665 when the University had to be closed because of the plague. Newton returned home and continued working and thinking intensely for the next few years.

Newton's own account of what took place in the next two years (taken here from R.S. Westfall's biography [3], see Box 22.1) leads to the conclusion that nobody has ever achieved so much in such a short span of time. Though some historians have raised doubts about Newton's 'recollections', it is clear that he did lay the foundations for new areas in mathematics, optics, and celestial dynamics at around this time. Interestingly, during the same period, Newton also developed a keen interest in alchemy, something he pursued vigorously for the rest of his life.

Box 22.1: The Closure of a University Helps Physics!

Here is Newton's own account of his ideas and work during the two years 1665–1666 (from Never at Rest, by R. S. Westfall, CUP) when the University was closed.

"In the beginning of the year 1665 I found the Method of approximating series & the Rule for reducing any dignity of any Binomial into such a series. The same year in May I found the method of Tangents of Gregory & Slusius, & in November had the direct method of fluxions & the next year in January had the Theory of Colours & in May following I had entrance into ye inverse method of fluxions. And the same year I began to think of gravity extending to ye orb of the Moon & (having found out how to estimate the force with wch [a] globe revolving within a sphere presses the surface of the sphere) from Keplers rule of the periodic times of the Planets being in sesquialterate proportion of their distances from the center of their Orbs, I deduced that the forces wch keep the Planets in their Orbs must [be] reciprocally as the squares of their distances from the centers about wch they revolve: & thereby compared the force requisite to keep the Moon in her Orb with the force of gravity at the surface of the earth, & found them answer pretty nearly. All this was in the two plague years of 1665–1666. For in those days I was in the prime of my age for invention & minded Mathematicks & Philosophy more then than at any time since."

There is one trait in Newton's character which has been a puzzle for historians. Newton harboured an abnormal fear of criticism and even a healthy academic debate of his ideas. Time and again he refused to let the world know of his work lest there should be adverse criticism and controversy. At the same time, he was very possessive of what he had discovered and could never share the credit for these discoveries with anyone else. Newton's scientific career could have been very different if he had published his results earlier or collaborated with other scientists; this might well have led to far greater scientific achievements on his part, and it would certainly have triggered a more rapid growth of science.

What finally prompted Newton to come out of his shell was the fact that, in 1668, there appeared a book titled *Logarithmotechnia* by Nicholas Mercator in which the author explained several results about infinite series. Newton had worked out the same results, in fact with greater generality, a few years earlier and was horrified to see someone else getting the credit. He hastily wrote a treatise called *De Analysi* and asked his friend, Isaac Barrow, at Cambridge to communicate it to a small group of London mathematicians. Even at this stage, Newton asked Barrow to withhold the name of the author of the book! Only when the treatise was received with a positive response did Newton reveal himself.

During the next few years, Newton polished up his early work on the branch of mathematics we now call calculus, and wrote it up as a short treatise. Though only a few knew it at that time, Newton had become the foremost mathematician of his time with this publication. Shortly thereafter, Isaac Barrow resigned from the Lucasian Chair at Cambridge and recommended Newton for the job. Newton was appointed the Lucasian professor of mathematics at the young age of 26 and held this position for the next 32 years.

During the first few years as Lucasian professor, Newton lectured on some of his works in optics. At that time, there existed two conflicting viewpoints regarding light. The original idea due to Aristotle treated light very qualitatively and considered phenomena like colours as arising out of the physical modification of light by materials. According to Aristotle, 'pure' light should be colourless and

Fig. 22.2: A drawing by Newton himself, describing his crucial experiment involving the prism. The first prism separates the white light into different colours and one of the colours, when passed through a second prism does not undergo any further change [10].

homogeneous. An alternative point of view, advocated by Descartes, treated all the optical phenomena as inherently mechanical. Descartes had made optics a quantitative science by, for example, stating clearly the laws governing reflection, refraction, etc. Newton adopted the mechanical view of light and, as usual, pushed it to its logical extreme. By sending a beam of white light through a prism and splitting it into various colours, Newton demonstrated that white light could be thought of as a heterogeneous mixture of different colours; what was more, he could show that when *one* of the colours is passed through a second prism, it does not undergo any further change (Fig. 22.2). He also showed that these colours could be mixed back together to produce white light again. In his view, material bodies only separated out the components which were present in white light, rather than modifying it. Using this idea, Newton could provide rather quantitative explanations for several optical phenomena, including the formation of the rainbow.

Since lenses and prisms always split white light into coloured bands, telescopes, which use lenses, suffer from a defect known as 'chromatic aberration'. This aberration causes coloured fringes to appear around the images seen through a telescope. Newton believed — though wrongly, because we now have achromatic lenses — that this defect could never be eliminated in telescopes using lenses. To tackle this problem, he developed the first reflecting telescope, which used concave *mirrors* rather than convex lenses (see Fig. 22.3). This telescope caused a sensation when it reached London in late 1671 and significantly helped in Newton's election to the Royal Society.

Newton presented his first paper on optics at the Royal Society in 1672 and a second in 1675. Both immediately came under attack from Robert Hooke (1635–1703), who was then one of the leading figures of the Royal Society [4]. Hooke wrote a condescending critique of the first paper and accused Newton of stealing his ideas in the second one. Newton, who could never respond to criticism rationally, was deeply annoyed and the two scientists became sworn enemies.

Around the same time, Newton was corresponding with a group of English Jesuits in Liege, who were also raising objections against Newton's theory of light. Their objections were rather shallow (and arose from a mistaken notion about Newton's experiments) but again Newton failed to react objectively. As a result, the correspondence dragged on for nearly three years and ended with Newton suffering a severe nervous breakdown in 1678. What was more, these bitter exchanges made Newton withdraw from the mainstream of intellectual life and become a recluse.

It was during these years that Newton turned to another passion of his — alchemy. He spent a significant amount of time copying ancient texts by hand and trying to make sense out of the mystical imageries present in them. Given this preoccupation, he failed to provide a scientific basis for chemistry — a task which was achieved by people of far lesser genius in the next century, showing that intellectual ability alone cannot precipitate a major scientific discovery. It is, however, possible that Newton's interest in alchemy had another favourable influence: it made him think in terms of 'attraction' and 'repulsion' between particles and the mechanical notion of the 'force' that particles exert on one another.

Around this time, Hooke tried to restart his correspondence with Newton on the topic of planetary motion, based on the notion that the planets are influenced by the force exerted by a central agency, possibly the Sun. During this correspondence — which was actually quite brief, with Newton terminating it abruptly — they debated the crucial question: What will be the path followed by a particle thrown from a tower on Earth? Newton actually drew a figure for this path as a spiral ending at the centre of the Earth. This was, of course, wrong and Hooke immediately pointed

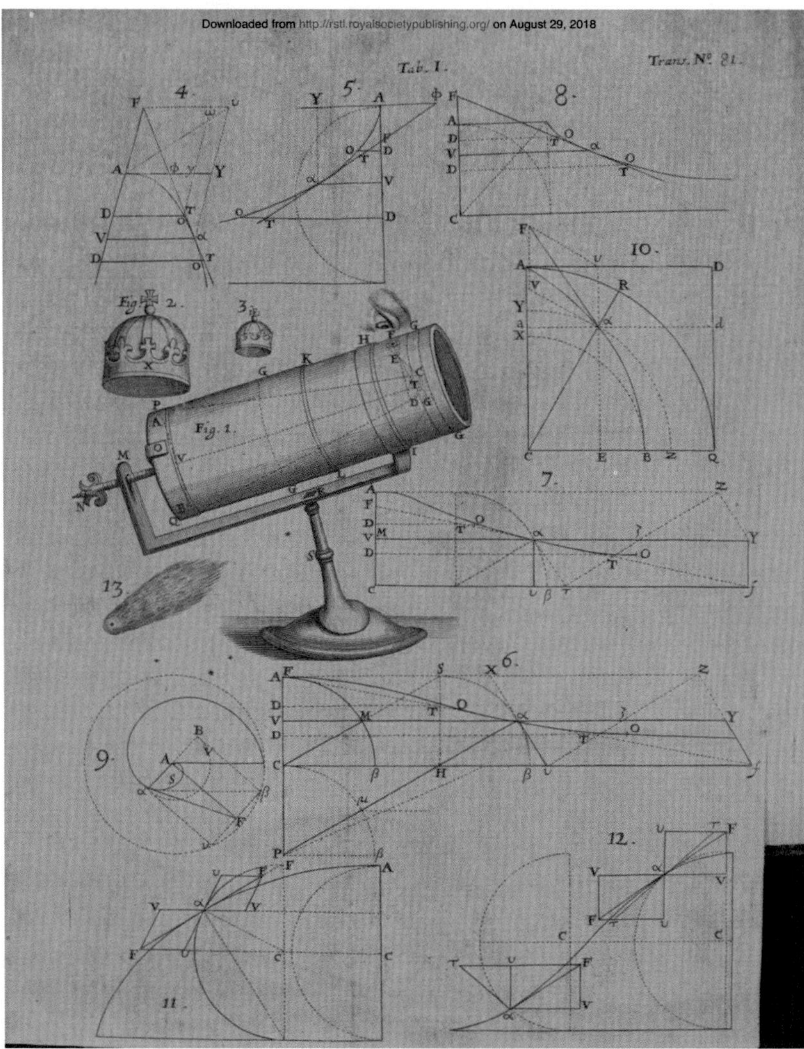

Fig. 22.3: The March 1672 issue of the Philosophical Transactions of the Royal Society had an account of Newton's telescope (marked Fig. 1 in the picture). The two crowns show how an object 300 feet away will look through the telescope made by Newton (marked 2) and through a more conventional 25-inch telescope (marked 3). Other pictures refer to various other articles appearing in that issue [11].

it out. According to Hooke, the path would have been elliptical and the particle would return to its original position if it could travel through the Earth. Newton hated being proved wrong (and especially by Hooke), but had to admit defeat this time. He, however, objected to the elliptical curve drawn by Hooke under the assumption that the gravitational force exerted by a body is a constant, independent of distance! Hooke immediately responded by saying that he assumed the force of gravity decreases as the square of the distance from the body.

Thereby hangs a tale of a bitter fight over priorities in the history of science. Hooke felt that he should also get credit as a co-discoverer of the (inverse-square) law of gravitation and Newton — quite characteristically — refused to share the prize of recognition. The turn of events which led to this historic controversy are as follows.

Sometime in 1684, Edmund Halley (1656–1742), the discoverer of Halley's comet, became interested in the problem of planetary orbits and asked Hooke whether he knew what force could cause the elliptical orbits. Hooke gave the correct answer but could not produce any detailed justification. Later on, in the same year, Halley visited Newton in Cambridge and asked him the same question. Newton not only gave the correct answer but also sent Halley a proof of this claim in the form of a short paper called *De motu corporum in gyrum* (On the Motion of Bodies in Orbits). It was the discussion with Halley which convinced Newton of the importance of his own work. By 1686, Newton developed — what started as a nine-page paper — the *De motu* into the classic work *Principia Mathematica*. He sent this work to the Royal Society for publication. The Society's finances were in poor shape at that time (partially due to the earlier publication of the handsome book *De Historia Piscium* — Latin for "The History of Fish" — which sold very poorly!); but Halley, who at that time was acting as publisher of the Philosophical Transactions of the Royal Society, decided to publish the book at his own expense. As soon as the *Principia* was submitted to the Royal Society for publication, Hooke raised a hue and cry and accused Newton of plagiarism. Newton's response was quite typical. He went through his manuscript and removed almost all the references he had originally made to Robert Hooke. Undoubtedly, the *Principia* was the work of an intellectual giant; but in his reaction to Robert Hooke, Newton also showed how small a man he was.

The *Principia* (Fig. 22.4) is made up of an introduction and three books. Together, they provide a systematic exposition of mechanics and a description of the system of the world. The introduction contains, among other things, the laws of motion, which allow mechanics to be formulated in a fairly rigorous manner. Books 1 and 2 consider different hypothetical forms of forces and the motion of

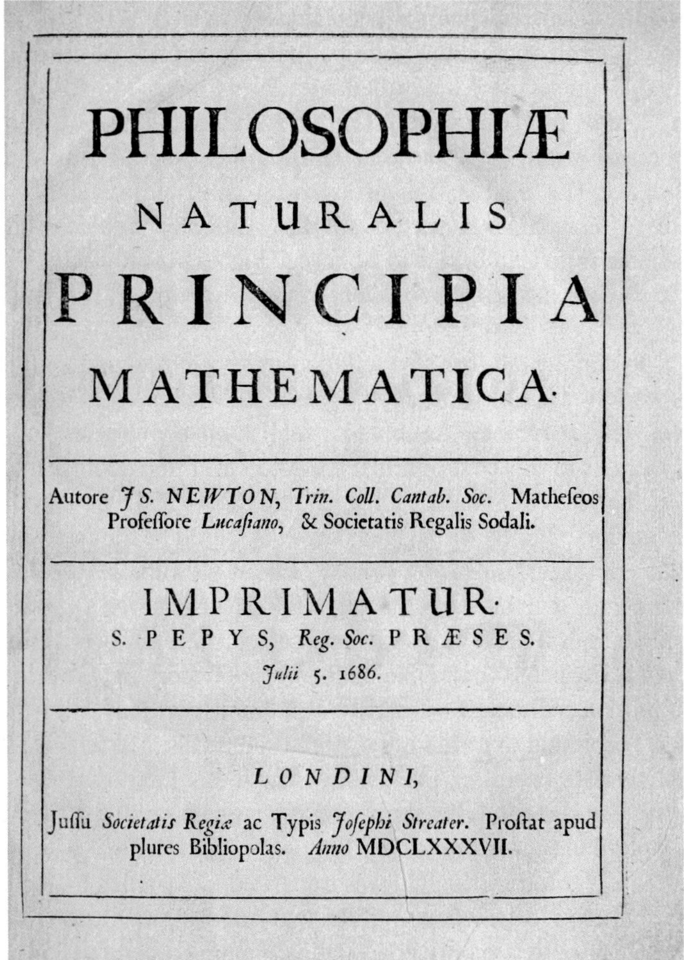

PHILOSOPHIÆ

NATURALIS

PRINCIPIA

MATHEMATICA.

Autore *J S.* NEWTON, *Trin. Coll. Cantab. Soc.* Matheſeos Profeſſore *Lucaſiano*, & Societatis Regalis Sodali.

IMPRIMATUR.

S. PEPYS, *Reg. Soc.* PRÆSES.

Julii 5. 1686.

LONDINI,

Juſſu *Societatis Regiæ* ac Typis *Joſephi Streater.* Proſtat apud plures Bibliopolas. *Anno* MDCLXXXVII.

Fig. 22.4: Title page [12] of Newton's magnum opus, *Principia*, first edition (1687).

particles under the action of these forces. Finally, Book 3 applies the general theory — developed in the previous two books — to the study of planetary motion. Most of the proofs use geometrical techniques, rather than algebraic or analytic methods, which was the tradition of those days. The third book also contains the statement of the law of universal gravitation and — for the first time in the history of science — a fundamental force was recognized and described as such.

The *Principia* resulted in international fame for Newton. Young British scientists took Newton as a role model and within about a generation, most of the salaried chairs in English universities were filled with 'Newtonians'. In 1689, Newton was elected the Member of Parliament for Cambridge University. However, he suffered a nervous breakdown in 1693 and regained stability only after a couple of years of convalescence. His creative life and scientific contributions slowly came to an end after this. Finally, in 1696, he took up the position of Warden of the Mint and moved residence from Cambridge to London. Honours continued to be heaped on him: in 1703, he was elected the president of the Royal Society and in 1705 he was knighted.

There is no doubt at all that Newton thoroughly enjoyed and took tremendous pride in his worldly success (which, of course, he fully deserved). On being knighted, for example, Newton took the trouble to establish his pedigree and applied to the College of Heralds for his coat of arms. Based on his often quoted statement, "If I have seen further, it is by standing on ye shoulders of giants", a sense of modesty is generally attributed to Newton. This, however, could be a misinterpretation. In his remarkable book, *On the Shoulders of Giants*, sociologist Robert Merton [5] argues convincingly that this particular expression had achieved a very conventional meaning by Newton's time. It was what the great and the noble people were expected to say on certain occasions (more like 'hello' or 'good morning') without, of course, meaning it in its true sense. So familiar was this usage, that themes based on it can be found decorating the windows of some cathedrals. Merton also lists several earlier and later uses of this expression by many other famous people. Of course, any objective historian studying the way Newton dealt with his fellow scientists would find it hard to accuse him of modesty!

Incidentally, Newton's Principia was translated into French by the lady Emilie du Chatelet (1706–1749), whose intellectual contributions to diverse areas of mechanistic philosophy are only now gradually being recognized. The complete translation appeared in 1759 and, for many years, remained the only version of the Principia in French. In addition to translating this work, she published her philosophical magnum opus, *Institutions de physique*, which contributed to the shift from Cartesian physics towards Newtonian physics in France.

Newton died in 1727 at the age of 85 — a death which triggered a display of pomp and pageantry, poems, statues, and other commemorations. He was buried at Westminster Abbey in the manner becoming of "a king who had done well by his subjects" as Voltaire put it.

Fig. 22.5: Portrait [13] of Leibniz (1646–1716), who developed the principles of calculus independently of Newton, this leading to another bitter feud over priorities. It is probable that Newton was the first to develop calculus, but Leibniz arrived at it completely independently [6] and published the results in 1684, before Newton did. The conduct of the controversy was far from gentlemanly, and Newton's behaviour was probably the worst. When he was the President of the Royal Society, Newton appointed an 'impartial' committee to investigate this issue, wrote the report, reviewed it 'anonymously' himself, and got it published by the Society [7]!

Box 22.2: Newton, Leibniz, and Calculus

One of the important mathematical contributions made by Newton was in systematizing the branch of mathematics we now call calculus, though, as we saw in a previous chapter, the basic ideas were developed in Kerala (India) centuries before. Newton put together several ideas which were known earlier and added some significant innovations of his own, thereby developing both differential and integral calculus. In the simplest terms, these topics deal with the manipulation of infinitesimally small quantities in a proper way (see Box 20.1). Newton's approach to calculus was essentially that of a mathematical physicist, and he thought of calculus as a calculational tool.

The principles of calculus were also developed independently in Europe by the German mathematician, G.W. Leibniz (1646–1716). As usual, Newton was dragged into another bitter feud over priorities — this time with national prides at stake [8]. It has now been established with reasonable certainty that Newton was indeed the first to develop calculus. However, Leibniz [6] arrived at it completely independently and published the results in 1684, before Newton did.

Though both men stooped very low in their conduct over the controversy, Newton's behaviour was probably the worst. At one stage in the debate, Leibniz made the mistake of appealing to the Royal Society to resolve the dispute. Newton, as president of Royal Society, appointed an 'impartial' committee made entirely of his friends to investigate the issue! Newton clandestinely wrote the report — officially accusing Leibniz of plagiarism — which was published by the Society, and even reviewed it 'anonymously' in the Royal Society's own periodical [7]. As a consequence of this dispute, British mathematicians were alienated from their European counterparts through most of the eighteenth century and, in fact, the quality of British mathematics fell behind that of continental Europe after Newton's death.

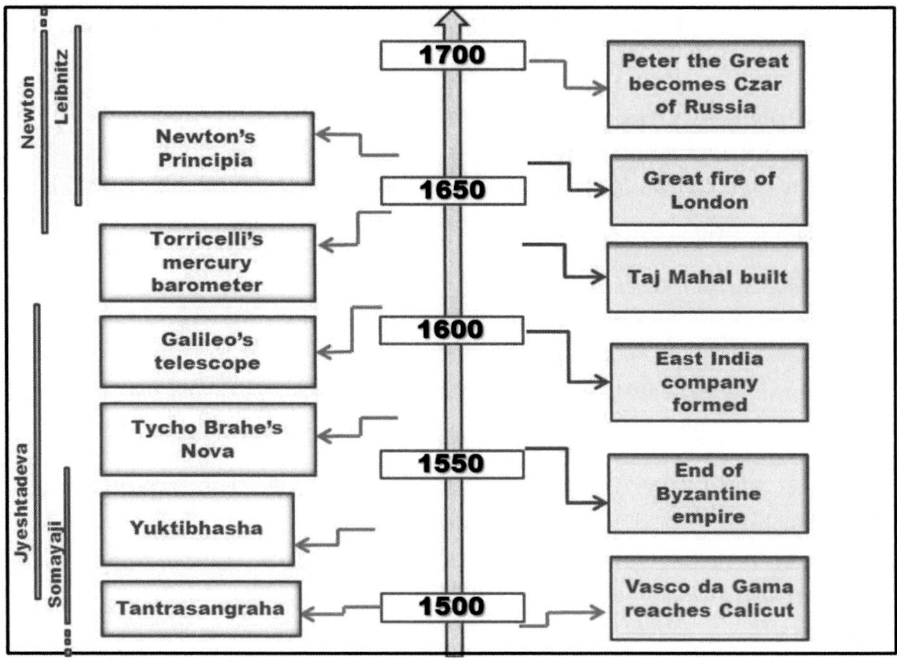

WHEN

Notes, References, and Credits

Notes and References

1. As we saw in Chap. 21, during Newton's lifetime, two calendar systems were in use in Europe: the Julian calendar in Protestant and Orthodox regions, including Britain, and the Gregorian calendar in Roman Catholic Europe. At the time of Newton's birth, Gregorian dates were ten days ahead of Julian dates: so he was born on 25 December 1642 in the Julian calendar, corresponding to the (modern) date of 4 January 1643. At the time of his death, the difference between the calendars had increased to eleven days: further, he died after the beginning of the Gregorian year on 1 January, but before that of the Julian new year on 25 March. He died on 20 March 1726 according to the Julian system, but the year is usually adjusted to 1727. A full conversion to the Gregorian system gives the date as 31 March 1727!
2. Here is a sample of literature dealing with different aspects of Newton:

Westfall, Richard S. (2007), *Isaac Newton*, Cambridge University Press, Cambridge [ISBN 978-0-19-921355-9].
Christianson, Gale E. (1996), *Isaac Newton and the Scientific Revolution*, Oxford University Press, Oxford [ISBN 0-19-530070-X].
Dobbs, B.J.T. (1983), *The Foundations of Newton's Alchemy or The Hunting of the Greene Lyon*, Cambridge University Press, Cambridge [ISBN 978-0521273817].
Gjertsen, Derek (1986), *The Newton Handbook*, Routledge, London, UK [ISBN 0-7102-0279-2].

3. Westfall, R. S. (1983), *Never at Rest: A Biography of Isaac Newton*, Cambridge University Press, Cambridge [ISBN 978-0521274357].
4. For more on Robert Hooke, see:
   Chapman, Allan (2004), *England's Leonardo: Robert Hooke and the Seventeenth-Century Scientific Revolution*, Institute of Physics Publishing, UK [ISBN 0-7503-0987-3].
   Inwood, Stephen (2002), *The Man Who Knew Too Much*, Pan, UK [ISBN 0-330-48829-5]. (Published in the US as *The Forgotten Genius*.)
5. Merton, Robert K. (1995), *On the Shoulders of Giants: A Shandean Postscript*, University of Chicago Press, Chicago, USA [ISBN-10: 0226520862].
6. See, e.g.,
   Aiton, Eric J. (1985), *Leibniz: A Biography*, CRC Press, USA [ISBN 978-0852744703].
   Antognazza Maria Rosa, (2008), *Leibniz: An Intellectual Biography*, Cambridge University Press, Cambridge [ISBN 978-0521806190].
   Jolley, Nicholas, (Ed.) (1994), *The Cambridge Companion to Leibniz*, Cambridge University Press, Cambridge [ISBN 978-0521367691].
7. See
   Hawking. S (1998), *A Brief History of Time*, Bantam, USA [ISBN 978-0553109535].
   http://hist.science.free.fr/storie/IERI/NewtonVoltaireEmilie/Newton%20vs_%20 Leibniz.htm
8. For more on the calculus controversy, see
   Brown, Richard C. (2012), *Tangled origins of the Leibnitzian Calculus: A case study of mathematical revolution*, World Scientific, Singapore [ISBN 9789814390804].
   Hall, A. R. (1980), *Philosophers at War: The Quarrel between Newton and Gottfried Leibniz*, Cambridge University Press, Cambridge [ISBN 978-0521227322].

Figure Credits

9. Figure 22.1 courtesy: Barrington Bramley [after Godfrey Kneller], Public domain, via Wikimedia Commons. https://commons.wikimedia.org/wiki/File:GodfreyKneller-IsaacNewton-1689.jpg (from public domain).

10. Figure 22.2 taken from:
    Fara P. (2015),"Newton shows the light: a commentary on Newton (1672) 'A letter [. . .] containing his new theory about light and colours [. . .]' ". Phil. Trans. R. Soc. A **373**, 20140213. http://dx.doi.org/10.1098/rsta.2014.0213 [dx.doi.org]
    and reproduced under the terms of the Creative Commons Attribution License http://creativecommons.org/licenses/by/4.0/: [creativecommons.org]
11. Figure 22.3 courtesy: From the Philosophical Transactions of the Royal Society, London; This figure is reproduced, with permission, from the journal Philosophical Transactions, 1672, 7, 0, published 1 January 1672 (Plate 1 of issue 81).
12. Figure 22.4 courtesy: The original uploader was Zhaladshar at English Wikisource [Public domain], via Wikimedia Commons.
    https://commons.wikimedia.org/wiki/File:Prinicipia-title.png (from public domain).
13. Figure 22.5 courtesy: Christoph Bernhard Francke [Public domain], via Wikimedia Commons.
    https://commons.wikimedia.org/wiki/File:Gottfried_Wilhelm_Leibniz,_Bernhard_Christoph_Francke.jpg (from public domain).

23 The Thirst for Power

Ancient civilizations used the muscle power of humans and animals in their agricultural and domestic pursuits. The efficiency, which could be achieved from such procedures, was necessarily quite small. Soon it was realized that water and wind could be harnessed to provide a more efficient source of power. Windmills were known to the Arabs fairly early on, while Europe seemed to have learnt about them only in the twelfth century — from the Crusaders who brought this knowhow home with them. Another source of power, tapped by all civilizations, was the water wheels installed in waterfalls and streams. Though windmills and water wheels could be considered as technological innovations, the real breakthrough came only with the use of steam. Indeed, the story of power from steam forms an interesting chapter in the history of technology.

The theoretical basis for using steam power came originally from the investigations of Otto von Guericke (1602–1686) and Robert Boyle (1627–1691). The practical necessity, on the other hand, arose from the rapid deforestation of England in the seventeenth century. The English Navy needed large amounts of wood for shipbuilding and, consequently, wood became too scarce to be used as a fuel. England, of course, had huge deposits of coal, which could serve as an alternative fuel. But the coal mines were repeatedly getting waterlogged, making them unusable for a good fraction of the time. The usual remedy was to pump the water out of the mines by hand or using horses, but this was quite complicated and slow.

It occurred to an English engineer, Thomas Savery (1650–1715), that the pressure of the air could be used to pump water out more efficiently [1]. The

T. Padmanabhan and V. Padmanabhan, *The Dawn of Science*,
https://doi.org/10.1007/978-3-030-17509-2_23

idea, essentially, was to fill a vessel with steam and then condense the steam; the vacuum produced inside the vessel would then suck the water up from the mine if a tube was connected between the vessel and the mine. This instrument, which was called the *Miner's Friend*, was in fact the first practical steam engine. Significant improvements in its design were later introduced by Thomas Newcomen (1663–1729), a blacksmith, who made very robust structures and carefully polished pistons. His improved machine (Fig. 23.1) was first installed in Staffordshire in 1712. Though the Newcomen engine served its purpose, it consumed far too much fuel. The next major breakthrough, which allowed for the economical use of the steam engine, came from James Watt (1736–1819), a Scottish engineer [2].

Watt was a sickly child and his childhood was rather uneventful. His father had a successful ship- and house-building business. Watt was tutored at home by his mother at first, but later joined a grammar school, where he learned the usual subjects: Latin, Greek, and mathematics. At the age of 17, he tried his hand as a mathematical instrument maker in Glasgow, and later, in 1755, in London. Returning to Glasgow a year later, essentially due to bad health, he opened a shop at the university in 1757, which made several mathematical instruments like quadrants, compasses, etc. There he met the Scottish chemist, Joseph Black (1728–1799), from whom he learnt some curious facts regarding heat energy. In particular, Black had conducted a series of experiments in thermodynamics which had made him realize that the quantity referred to as *heat energy* was not the same as the *temperature*.

For instance, when one heats ice, it absorbs the heat energy and melts, even though its temperature does not change. Similarly, he noticed that the energy supplied to water at its boiling point went into converting water to steam without changing its temperature. These observations implied that there was more heat content in steam at 100°C than in boiling water at the same temperature. Black named this heat content 'latent heat' and mentioned these results to James Watt.

This knowledge was of crucial importance for James Watt. In 1764, he was called upon to repair a Newcomen steam engine and while doing so, he realized that it was the latent heat that was causing the most significant wastage of energy in these engines. To condense the steam, the vessel which contained the steam had to be cooled; but then it had to be refilled with steam for the next cycle of operation. Most of the energy now went into merely heating the chamber back up to a high temperature. Thus, in every cycle of operation, a large amount of energy was wasted because the vessel was repeatedly heated and cooled.

Watt came up with an ingeniously simple solution to this problem. He introduced a second chamber (now called the 'condenser') into which the steam could be fed. It was now possible to keep the first chamber (called the 'cylinder') permanently hot

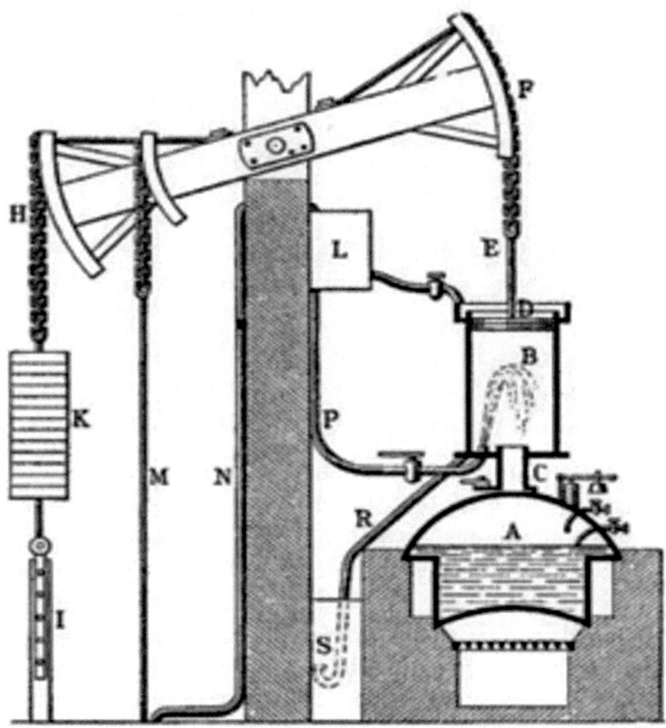

Fig. 23.1: The Newcomen steam engine, designed by Thomas Newcomen, was a precursor to more modern steam engines [4]. The first such machine was installed in Staffordshire in 1712. The improved design by James Watt replaced the Newcomen engines almost completely.

and the second chamber permanently cold, thereby avoiding the problem of heating and cooling the chamber. Shortly afterwards, he met John Roebuck (1718–1794), a British chemist, inventor, and industrialist, and founder of the Carron Works (which manufactured iron). Roebuck urged Watt to make a prototype of the engine. They entered into a partnership in 1768 and in the following year, Watt took a patent for "A New Invented Method of Lessening the Consumption of Steam and Fuel in Fire Engines".

The practical difficulty in implementing Watt's design on a commercial basis was in machining the piston and cylinder to the perfection that was required.

Fig. 23.2: James Watt (1736–1819), who significantly improved on the earlier design of the Newcomen steam engine [5]. By 1769, Watt had produced a practical steam engine with far greater fuel efficiency than the earlier Newcomen engines. In the next few years, Watt introduced further crucial improvements, and by 1800, his engines had totally replaced the older Newcomen engines with hundreds of these working in England. (But Watt did not *invent* the steam engine, as is sometimes claimed.) The consequences of this single invention were momentous. Large-scale production of commodities became cheap, handicrafts became commercially unviable, and the artisans were replaced by factory workers. The industrial revolution, one could say, was triggered by the steam engine.

Workers in the iron factories were more blacksmiths than machinists in those days, and they sometimes found it difficult to produce items with the necessary precision. Watt also had to spend a fair amount of his resources in obtaining a patent for the invention. In fact, this lack of resources forced Watt to take up employment — first as a surveyor, then as a civil engineer — for eight years!

All this prevented further progress for a while. Added to this, Roebuck went bankrupt in 1772 and Watt needed to look elsewhere for support. Around 1774, Watt moved to Birmingham and soon found a working partnership with the English manufacturer Matthew Boulton (1728–1809), who operated the Soho Works in Birmingham. Boulton, of course, took a share in Watt's patent but successfully campaigned for it to be extended by an act of Parliament. Very soon, persistence paid off and Watt's dream child became a commercial success. His steam engine had far greater fuel efficiency than a Newcomen engine and it worked considerably faster, since there was no need to pause between heating and cooling the chambers.

The very first engines were all used for powering the pumps that removed water from mines. The design was commercially successful, and for the next five years or so, Watt was busy installing more engines, especially in Cornwall, for pumping water out of the mines.

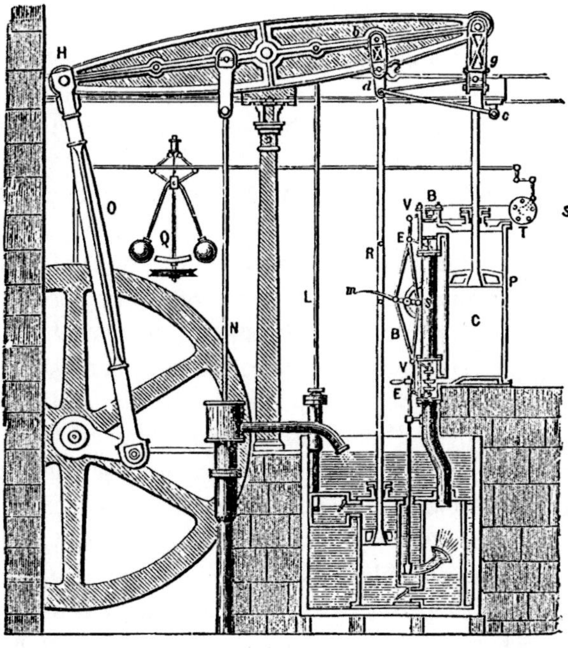

Fig. 23.3: One of the first steam engines based on the double-acting principle, developed by Mathew Boulton and James Watt [6].

In the following years, Watt introduced further improvements in the design which made the steam engine truly versatile (Fig. 23.3). For example, he made arrangements for the steam to enter alternately from both sides of a piston, so that work could be extracted during both the pushing and pulling motions of the piston; this innovation nearly doubled its performance. He also devised mechanical attachments which converted the back-and-forth movements of the piston into a rotary movement of the wheel. The steam engine thus became the first modern device which could be applied for different purposes, by using the energy that occurred in nature (in the form of fuel) to run virtually any form of machinery.

By 1800, his engines had totally replaced the older Newcomen engines, with over 500 of these working in England. They were used in paper mills, flour mills, cotton mills, iron mills, distilleries, and waterworks. By 1790, Watt had earned more than 76 000 pounds in royalties on his patents (in just about a decade) and had become a very rich man. This meant he was often involved in lawsuits to protect his patents.

Just thirty years after Watt invented the steam engine, steam power was used to make the first railway engine, and this was used for transporting coal. In 1829, George Stephenson won a competition, building an engine that could pull passenger trains at 50 km per hour. While it took about 12 days to travel from Edinburgh to London on horseback, a train driven by a steam engine did it in just 8 hours. England soon became the centre for manufacturing trains.

The consequences that came from tapping the power of steam, by Watt and others who followed him, were far-reaching. Steam engines, powered by coal, could deliver energy constantly at any location, which meant that manufacturing locations and factories no longer needed to be near streams or waterfalls. Large-scale production of commodities became cheap with the availability of comparatively unlimited sources of power. Handicrafts became commercially unviable and the artisan was replaced by the factory worker. Cities grew along with industries, and so did urban life and all the benefits and drawbacks of the factory system. In short, the industrial revolution was triggered by the steam engine. One example was the mechanization of the textile industry, which was of prime importance to England. Richard Arkwright (1732–1792) and others invented machines which replaced labour in textile manufacturing, thereby becoming the first group of 'capitalists'.

Box 23.1: Patently Unjust?

The patents issued to Watt and Boulton gave them a virtual monopoly in the manufacture of steam engines. It has been argued by social scientists and historians that it actually harmed the cause of technological development. Boulton and Watt could eliminate and suppress competition using legal means, granted by the patent, rather than having to improve their design and sell the engines more cheaply, as would have happened in open market competition [3].

One case which is often cited to prove this claim is that of an inventor Jonathan Hornblower (1753–1815) who designed what is called a compound steam engine, and even patented it in 1781. It had two cylinders, instead of one as in Watt's original model, and used high pressure steam. In spite of the improved efficiency and innovative technology, Hornblower was, unfortunately, blocked from commercializing his idea by Watt and Boulton, through litigation. The main point of the contention was the use of a condenser, even though the other innovative aspects played an equally crucial role in the design.

It has been argued on the basis of such examples that innovation was stifled until Watt's patents expired in 1800. Innovative designs for steam engines, irrespective of how much better they were compared to Watt's, had to use the idea of a separate condenser. The 1775 patent essentially provided Boulton and Watt with a monopoly over that idea, and hence several other improvements of key economic and technological value could not be introduced by others. It seems that many competitors just shelved their ideas until Watt's patent expired rather than face a legal wrangle.

In fact, the fuel efficiency of steam engines changed very little during the period of validity of Watt's patent, but then, between 1810 and 1835, it increased by almost a factor of five!

James Watt's work was duly recognized in his time. He was elected Fellow of the Royal Society in 1785 and became a foreign associate of the French Academy of Sciences in 1814. He also earned considerable wealth from his royalties. In one of his early experiments, Watt noticed that a strong horse could raise a weight of a hundred pounds nearly four feet in about a second and coined the term 'horsepower', defining it as 550 foot-pounds per second. Today, however, the metric system measures power using the unit 'watt' in honour of this inventor of the steam engine, with one horsepower being equal to 746 watts.

Watt died in 1819 in the attic of his home, Heathfield, outside Birmingham. His workshop is now preserved in the Science Museum, London, with all the furniture and instruments essentially as they were left at the time of his death.

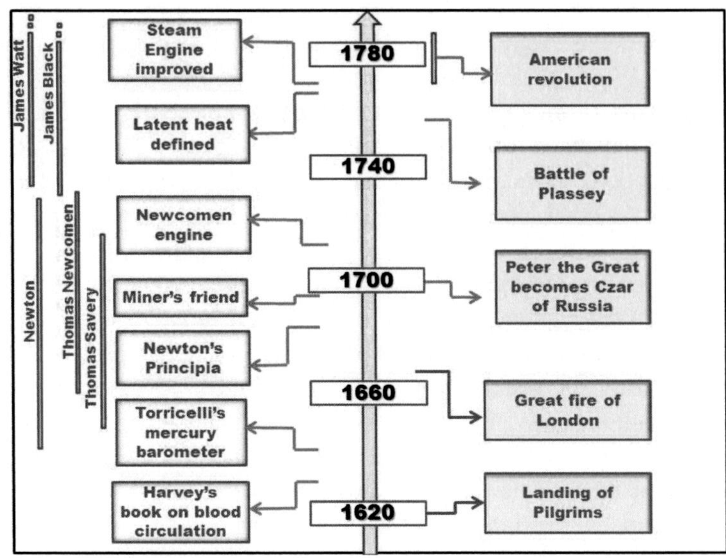

WHEN

Notes, References, and Credits

Notes and References

1. For more on these developments, see:
   Jenkins, Rhys (1936), *Savery, Newcomen and the Early History of the Steam Engine in The Collected Papers of Rhys Jenkins*, Newcomen Society, Cambridge [ASIN: B0012J2786].
   Rolt, Lionel Thomas Caswell (1963), *Thomas Newcomen: The Prehistory of the Steam Engine*, David and Charles, Dawlish [ASIN: B0000CLQ7F].
   Rolt, Lionel Thomas Caswell and Allen, John S. (1977), *The Steam Engines of Thomas Newcomen*, Moorland Publishing Company, Hartington [ISBN 0-903485-42-7].
2. For more on James Watt, see:
   Dickinson, H. W. (2010), *James Watt: Craftsman and Engineer*, Cambridge University Press, Cambridge [ISBN 978-1108012232].
   Hills, Richard L. (2002), *James Watt, Vol 1, His time in Scotland, 1736–1774; Vol 2, The years of toil, 1775–1785; Vol 3, Triumph through adversity, 1785–1819*, Landmark Publishing Ltd [ISBN 1-84306-045-0].

Marsden, Ben (2002), *Watt's Perfect Engine*, Columbia University Press, New York [ISBN 0-231-13172-0].

3. See, e.g.,
Boldrin, Michele and Levine, David K. (2008), *Against Intellectual Monopoly*, Cambridge University Press, Cambridge, UK [ISBN-10: 0521879280].
For an opposite point of view, see, e.g.,
Selgin, George and Turner, John L. (2010), *Strong Steam, Weak Patents, or, the Myth of Watt's Innovation-Blocking Monopoly, Exploded*, The Journal of Law and Economics, University of Chicago Press, Vol. **54** (4), pp. 841–861 [DOI: 10.1086/658495].

Figure Credits

4. Figure 23.1 courtesy: Meyers Konversationslexikon 1890 [Public domain], via Wikimedia Commons.
https://commons.wikimedia.org/wiki/File:Newcomens_Dampfmaschine_aus_Meyers_1890.png (from public domain).
5. Figure 23.2 courtesy: Antonia Reeve, [Public domain], via Wikimedia Commons.
https://en.wikipedia.org/wiki/File:James-watt-1736-1819-engineer-inventor-of-the-stea.jpg (from public domain).
6. Figure 23.3 courtesy: By Robert Henry Thurston (1839 – 1903) [Public domain], via Wikimedia Commons.
https://commons.wikimedia.org/wiki/File:SteamEngine_Boulton%26Watt_1784.png (from public domain).

24

Chemistry Comes of Age

Yet another branch of endeavour to come of age in the eighteenth century was chemistry. What helped it to grow into an exact science was the work of Henry Cavendish (1731–1810) and Joseph Priestley (1733–1804) from England, with extensive contributions by Antoine Lavoisier (1743–1794) from France.

Cavendish, a descendant [1] of two prestigious families of Dukes, had his early education in London, followed by four years at Peterhouse College in Cambridge. Though he completed his studies, he never took the final degree — for reasons which are rather unclear. After a short tour of the continent, he settled in London in 1755 and started working with his father, who was a skilled experimentalist. While he started out as an assistant to his father, Cavendish was soon designing his own experiments and breaking new ground, especially in the study of properties of gases and electricity.

When he was around 40 years old, Cavendish inherited a considerable fortune from a relative and found himself to be a millionaire. (This led a contemporary scientist, Jean-Baptiste Biot (1774–1862), to remark that Cavendish was the richest of all the learned men and the most learned of all the rich!) Actually, Cavendish was the archetype of the eccentric genius. He dressed shabbily — he usually wore a crumpled and faded suit and a three-coloured hat — spoke hesitantly and as little as possible, never appeared in public, and could not stand even the sight of women. So much so that he communicated with his (female) housekeeper by daily notes and ordered all female domestics to keep out of his sight! He disdained public acclaim;

T. Padmanabhan and V. Padmanabhan, *The Dawn of Science*,
https://doi.org/10.1007/978-3-030-17509-2_24

Fig. 24.1: Henry Cavendish (1731–1810) was a brilliant scientist who could justify being called an eccentric genius [4]. He was extremely shy, dressed shabbily, spoke very little, and could not stand the sight of women. He had become a millionaire by inheriting a fortune from a relative and was considered to be 'the richest of all the learned men and the most learned of all the rich!' In the field of chemistry, he isolated hydrogen and studied several of its properties; in particular, he produced water by injecting an electric spark to explode a mixture of hydrogen and air. This result was of crucial significance in the history of chemistry; it showed that one of the five elements the world was supposed to be made of, viz., water, could actually be formed by combining two of the other substances. In physics, he developed — what we now call — the Cavendish balance, a valuable tool for measuring very small forces. This allowed him to measure Newton's gravitational constant, which occurs in the law of gravitational attraction, and thus to weigh the Earth indirectly.

but he *did* accept fellowships of the Royal Society and the Institute of France. He rarely published his results and consequently slowed down the march of science.

In terms of scientific calibre, however, Cavendish was indeed brilliant. In 1766, he communicated to the Royal Society some early results describing his work on an inflammable gas produced by the action of acids on metals. Though this gas had been noticed before, Cavendish was the first to investigate its properties systematically. (Twenty years later, this gas was named hydrogen by Lavoisier.) Cavendish prepared the gas by the action of acids on various metals. He used hydrochloric acid (known as spirit of salt) and dilute sulfuric acid (known as oil of vitriol) and studied three metals: zinc, iron, and tin. He collected the gas by displacing water, which was an ingenious way to capture and store gases released during a chemical reaction (Fig. 24.2). This enabled Cavendish to measure properties like volumes and weights very precisely.

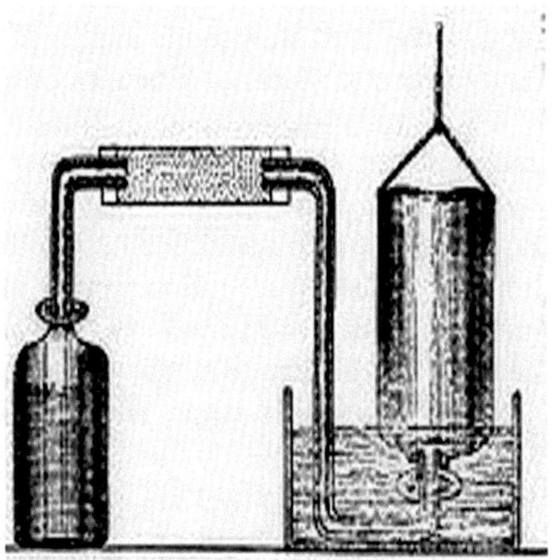

Fig. 24.2: Cavendish's apparatus for producing and collecting hydrogen [5]. Metals like zinc were placed in the bottle on the left and made to interact with acids. A bent glass tube was inserted into it and its other end was placed inside a vessel containing water. He then filled another ('collecting') bottle with water and inverted it over the end of the bent tube in the vessel of water. The collecting bottle was suspended by means of a string. This was an ingenious procedure for collecting gases that were insoluble in water (or using mercury instead of water to collect gases that were soluble in water) and allowed Cavendish to make very precise measurements.

By careful measurements, Cavendish was able to determine the density of hydrogen and discovered that it was significantly lighter than air. One of the popular concepts in chemistry in those days was that of a hypothetical substance called phlogiston (which, in Greek, means 'to set on fire'). All combustible substances were supposed to contain large amounts of phlogiston, and the process of combustion was thought to involve the loss of phlogiston. For example, wood was supposed to contain a lot of phlogiston, but not ash, and this was supposed to explain the fact that wood can burn but ash cannot. The remarkable lightness of hydrogen — and the fact that it helped combustion — made Cavendish conclude (erroneously, of course) that he had actually isolated phlogiston. It took several more years for the concept of phlogiston to be laid to rest.

However, Cavendish noticed that, when a mixture of hydrogen and air was exploded by injection of an electric spark, water was produced. Though similar experiments had been performed earlier by Priestley and even by James Watt, they do not seem to have grasped its significance. This result was of crucial conceptual importance because it gave the final blow to the medieval idea that water was a pure element; it was now clear that water could be formed by a suitable chemical reaction.

Cavendish also used electric sparks to make nitrogen combine with air (actually with the oxygen in air, to use modern terminology) forming an oxide, which he dissolved in water to produce nitric acid. He kept adding more air into the reaction expecting to use up all the nitrogen. However, he discovered that a small bubble of gas, amounting to less than one per cent of the whole, remained without participating in the reaction. This led him to speculate, quite correctly, that normal air contained a small quantity of a very inert gas. We now know that component to be essentially argon.

In addition to his contributions to chemistry, Cavendish also made significant contributions to the study of electrical phenomena and to the measurement of the gravitational constant (Box 24.1). Cavendish died at the age of 78. He left his large fortune to his relatives and almost nothing to science. This omission was later rectified by the Cavendish family in 1875, when they set up the Cavendish laboratory at Cambridge University, which contributed tremendously to the development of science in the next century.

Box 24.1: Weighing the Earth

According to Newton's law of gravitation, two bodies of masses M and m, when separated by a distance r, attract each other with a force given by $F = GMm/r^2$, where G is the gravitational constant. It follows from this law that G can be determined if the force of attraction between any two bodies can be measured accurately enough. Once G is known, it is possible to determine the mass of the Earth using the known value of the force by which Earth attracts all bodies.

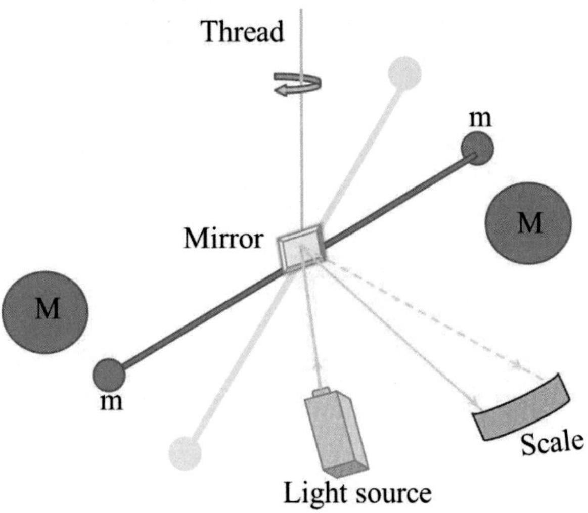

The difficulty, of course, is that the gravitational force F is extremely tiny at laboratory scales and requires a very sensitive piece of apparatus to measure it. One of Cavendish's contributions was to design a suitable setup (see figure above). This consisted of a light rod suspended from the middle by a wire. At each end of the rod was a light ball made of lead. The rod could twist freely about the wire and even a small force applied to the lead balls would make it twist. Cavendish brought two other large balls near the light balls, one on either side. By measuring the twist produced on the rod, he could calculate the force of attraction between the balls and hence estimate the constant G. From the estimated value of G, he could calculate the mass of the Earth to be 6.6×10^{21} tons and its density to be about 5.5 g/cubic centimetre. The Cavendish balance remains a valuable tool in the measurement of small forces even today.

Another English chemist who lived around the same period [2] was Joseph Priestley (1733–1804). He was the son of a non-conformist preacher and was quite radical in his views, about both religion and politics. In his early days he studied languages, logic, and philosophy and not much science. He first took up

Fig. 24.3: Joseph Priestley (1733–1804) started out as a school teacher in Cheshire and authored several books on English grammar, education, and history [6]. However, he was very curious about science and kept in touch with scientists by spending about a month each year in London from 1765 onwards. He soon turned his attention to chemistry and ended up discovering oxygen, though he probably did not quite appreciate the importance of what he had done. But when Lavoisier heard about the discovery, he immediately recognized it as a new gas and named it oxygen. Priestley, being a unitarian by faith, had rejected several fundamental doctrines of Christianity. He was also a strong supporter of the principles which inspired the French Revolution, thereby making himself quite unpopular in England. On 14 July 1791, during the violent clashes between the local public of Birmingham and the supporters of the French Revolution (who attempted to celebrate the second anniversary of the fall of the Bastille), Priestley's house, laboratory, and library were burnt down. Priestley just about managed to escape to London. Later on, he migrated to America.

a job as a teacher in a day school in Cheshire and during this period, he actually wrote several books on English grammar, education, and history! Though he never studied science formally, he was curious about several new discoveries occurring around him at the time. Because of this interest, from 1765 onwards, he made a point of spending one month every year in London, where he could keep in touch with leading scientists. Strongly influenced by Benjamin Franklin (1706–1790), he wrote a book, *The History and Present State of Electricity*, which immediately earned him a place among scholars.

Soon Priestley's curiosity migrated from physics to chemistry, with a focus on the behaviour of gases. Only three distinct gases were known at that time — air, carbon dioxide (which was discovered by Joseph Black), and hydrogen (which Cavendish had just discovered). Priestley went on to isolate and study several more gases — such as ammonia and hydrogen chloride — from suitable reactions. He used a new technique of collecting the gas over mercury (rather than over water, which was the existing practice). In this way, he could collect many gases which were soluble in water.

His major discovery came in 1774. It was known that mercury, when heated in air, would form a brick-red coloured 'calx' (which we now know is mercuric oxide). Priestley found that when calx was collected and heated in a test tube, it turned back into mercury again, but let out a gas with interesting properties. Combustibles burned more brilliantly and rapidly in this gas, mice were particularly frisky in this atmosphere, and he himself felt 'light and easy' when he breathed it. Since Priestley believed in the prevailing phlogiston theory, he reasoned that the new gas must be particularly poor in phlogiston. So he called it 'dephlogisticated' air; but when Lavoisier (1743–1794) heard about the discovery he immediately recognized it for what it was, viz., a new gas, and christened it oxygen.

During another one of his experiments, Priestley dissolved carbon dioxide in water and found that the solution was a pleasant and refreshing drink. Though he didn't make a commercial success of it, he deserves to be called the father of the soft drink industry!

In 1779, Priestley moved to Birmingham as minister of the New Meeting Congregation. Being a Unitarian by faith, he had rejected most of the fundamental doctrines of Christianity, including the Trinity, predestination, and the divine inspiration of the Bible. In addition, he was a strong supporter and defender of the principles which inspired the French Revolution. These facts — and his publications on these subjects — made Priestley extremely unpopular with the local community. So, on 14 July 1791, the second anniversary of the fall of the Bastille, when the supporters of the French Revolution organized a meeting in Birmingham, the general public was not sympathetic, and decided to teach a lesson to those who had gathered. In the mob violence that followed, Priestley's house, laboratory, and library were burnt down and Priestley only just managed to escape to London.

However, the progress of the French Revolution, the execution of Louis XVI in France, and the declaration of war between France and Britain made life more and more difficult for Priestley. In 1794, he left Britain forever and migrated to the United States where he spent his last ten years. He was probably the first scientist to emigrate to the USA to escape local persecution, and certainly not the last! To

Fig. 24.4: Priestley Medal [7]. In 1794, Priestley migrated to the United States where he spent the last ten years of his life. To commemorate Priestley's scientific achievements, the American Chemical Society introduced the Priestley Medal in 1922 as its highest honor.

commemorate Priestley's scientific achievements, the American Chemical Society named its highest honour the Priestley Medal in 1922.

While both Cavendish and Priestley were outstanding scientists of that age, they had a contemporary, Antoine Laurent Lavoisier (1743–1794) who outshone them both [3]. Lavoisier is acclaimed as the father of modern chemistry and is sometimes called the 'Newton of chemistry', quite deservedly. But, in spite of his intellectual prowess, his life ended in a tragedy (Box 24.2); if Priestley suffered for supporting the French Revolution while living in the wrong country, Lavoisier lost his life for being in the wrong camp at the time of the French Revolution.

Lavoisier was born in Paris in a well-to-do family and had a wonderful childhood and an excellent education. He was a brilliant student and after dabbling a little in geology, turned his attention to chemistry where he made his mark. There are several contributions which Lavoisier made to this subject, any one of which could have been called a milestone in the history of science.

To begin with, Lavoisier championed the need for precise measurements to settle issues in chemistry, and in this respect, he did for chemistry what one might say Galileo did for physics. For example, there were people who believed at that time that water could be turned into earth (that is, one of the five elements could be converted into another) by a long period of heating. This was based on the fact that water, heated in vessels for many days, *did* develop a solid sediment. Lavoisier put this idea to test in 1786 by boiling water for about a hundred days in controlled conditions and weighing both the water and the vessel before and after the heating. The sediment did appear but Lavoisier could show conclusively — thanks to his precise measurements — that the weight of the water did not change. However, the vessel lost precisely the amount of weight equal to that of the sediment, thereby clearly showing that the sediment came from the vessel and not from the water.

A second major contribution of Lavoisier was to kill the phlogiston theory of combustion, which had been in vogue for nearly a century, even though it could not explain very much. Lavoisier found that, when he burnt phosphorous and sulphur, the final products weighed *more* than the original ones, clearly showing that some material had been *gained* from the air. This was a weight gain which the phlogiston theory could not explain at all. To test this point further, in 1774, he performed the experiments again in controlled conditions. This time he heated tin and lead in closed containers with a fixed supply of air. Both metals produced a layer of calx on the surface. The weight of the calx was indeed greater than the weight of the metal it replaced, but the entire system (made of metal, calx, air, vessel, etc.) was no heavier after the heating than before. He also demonstrated that the gain in weight did involve the loss of some amount of air, thereby creating a partial vacuum within the container. This conclusively proved that rusting and combustion did not involve a loss of phlogiston but a gain of at least some part of the air. This killed the phlogiston theory and established the modern basis for theoretical chemistry. It also led to the law of conservation of mass, which became a guiding principle in chemistry in the years to follow.

In October 1774, Priestley went to Paris to discuss his experiments with Lavoisier, and in particular, what he considered to be the dephlogisticated air. Lavoisier understood immediately that this notion was nonsensical and arrived at the correct interpretation: the air contained two gases, one of which supported combustion (which he called oxygen) and the other, which he called azote (from the Greek word meaning ‘no light’). Later on, in 1790, it was named nitrogen by Jean-Antoine Chaptal (1756–1832) and that is the name we now know it by.

Lavoisier also removed the phlogiston from the results of Cavendish. In 1783, Cavendish showed that water could be formed by taking the inflammable gas he had discovered and burning it in air. Lavoisier repeated the experiments in an improved manner and named the inflammable gas hydrogen (from the Greek words meaning 'to give rise to water'). All this fitted with his ideas on chemistry and the systematic approach he had been advocating.

Another notable contribution by Lavoisier was to streamline the principles by which every chemical substance is assigned a name, based on the elements making it up. In 1787 Lavoisier co-authored a book [along with Berthollet (1748–1822) and Fourcroy (1755–1809)] with the title *Methods of Chemical Nomenclature*, in which this system was established. It was so clear and logical that it was almost immediately adopted by almost all chemists everywhere.

One blemish that historians point out in Lavoisier's character was that he never gave credit where credit was due. For example, in the case of oxygen, Lavoisier avoided mentioning the help he had received from Priestley and did his best to give the impression that he, Lavoisier, had himself discovered oxygen. It is true that Lavoisier deserved full marks for everything related to oxygen *except* its discovery. However, the discovery of an element was what Lavoisier coveted. While he had done more for chemistry than any man before, it was not his destiny to actually discover any element! In the case of Cavendish, Lavoisier also tried to create an impression that the experiment of burning hydrogen in air was his own original idea. This again earned Lavoisier a dubious reputation as a credit snatcher.

Ironically, while Lavoisier heralded a revolution in chemistry, his life was put to an end by the other revolution in his country — the French Revolution, which unleashed a reign of terror costing many lives (Box 24.2). Lavoisier was branded a traitor because of his indirect involvement with a private taxation firm called Ferme Generale. He was convicted and guillotined in Paris on 8 May 1794, at the age of 50. His execution shook French science, making the French mathematician Lagrange comment: "A moment was all that was necessary to strike off his head; but probably a hundred years will not be sufficient to produce another like it."

Box 24.2: Guillotining Science

Lavoisier committed two costly mistakes which led to his tragic end. First, he invested half a million francs in the Ferme Generale in 1768 to earn money for his research. This private firm was engaged by the French government to collect taxes. The deal was that they would pay the government a fixed fee, but anything over and above that amount which they collected would be theirs. Needless to say, these 'tax farmers' were the most hated group in eighteenth century France. While Lavoisier did not himself engage actively in tax collection, he *did* earn nearly 100 000 francs a year out of this. Almost all of his earnings were used to set up a magnificent private laboratory for research in chemistry which was visited, for example, by Thomas Jefferson and Benjamin Franklin. All altruism aside, Lavoisier *did* earn money from tax farming.

Second, in 1771, he married Marie-Anne, who was the daughter of a key executive of the Ferme Generale. She was an intelligent and devoted wife and it was a splendid match but, alas, she *was* the daughter of a tax farmer.

By 1792, the radical anti-monarchists were in control in France and the tax farmers were hunted down. Very soon Lavoisier was arrested and tried. It is claimed that one Jean Pal Marat, a journalist who thought of himself as a scientist, accused Lavoisier of all sorts of ridiculous plots (including "adding water to people's tobacco") and went to great lengths to get Lavoisier condemned to death. This has been attributed to Marat's personal animosity toward Lavoisier. It is rumoured that Marat had ambitions of getting elected to the French Academy of Sciences, which were thwarted — for very good reasons — in 1780, by Lavoisier. In fact, Marat himself was assassinated in July 1793, but the verdict had already been reached. Lavoisier, along with his father-in-law and other tax farmers, was guillotined on 8 May 1794 and buried in an unmarked grave — just two months before the radicals were overthrown. About his death, Lagrange made the famous remark: "A moment was all that was necessary to strike off his head; but probably a hundred years will not be sufficient to produce another like it."

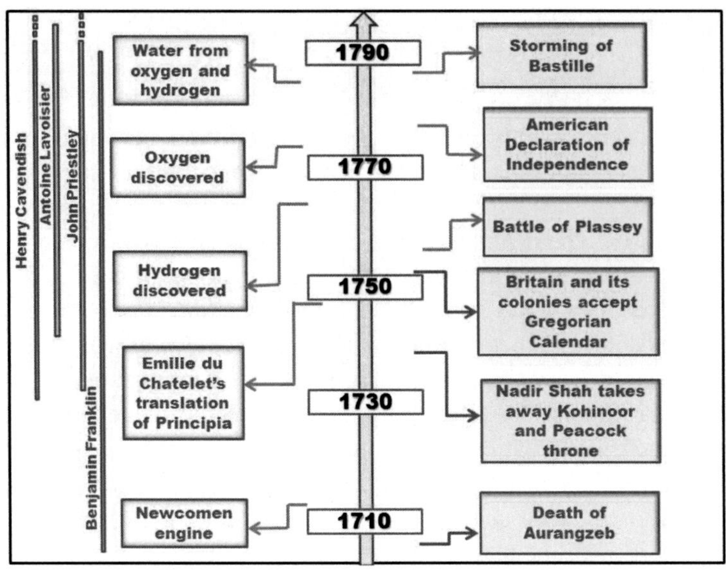

WHEN

Notes, References, and Credits

Notes and References

1. For more on Cavendish, see:
   Jungnickel, Christa and McCormmach, Russell (1999), *Cavendish: The Experimental Life*, Bucknell University Press, USA [ISBN 0-8387-5445-7].
   Moore, F. J. (1939), *A History of Chemistry*, McGraw-Hill, New York [ASIN: B002CERX6Q].
2. For more on Priestley, see:
   Gibbs, F. W. (1965), *Joseph Priestley: Adventurer in Science and Champion of Truth*, Thomas Nelson and Sons, London [ASIN: B00CJ9I4H6].
   Holt, Anne (1931), *A Life of Joseph Priestley*, Oxford University Press, London [ASIN: B0006ALE6I].
   Jackson, Joe (2005), *A World on Fire: A Heretic, an Aristocrat and the Race to Discover Oxygen*, Viking, New York [ISBN 0-670-03434-7].
   Johnson, Steven (2008), *The Invention of Air: A Story of Science, Faith, Revolution, and the Birth of America*, Riverhead, New York [ISBN 1-59448-852-5].

Schofield, Robert E (1997), *The Enlightenment of Joseph Priestley: A Study of his Life and Work from 1733 to 1773*, Pennsylvania State University Press, University Park Pennsylvania [ISBN 0-271-01662-0].
Uglow, Jenny (2002), *The Lunar Men: Five Friends Whose Curiosity Changed the World*, Farrar, Straus and Giroux, New York [ISBN 0-374-19440-8].

3. For more on the life and times of Lavoisier, see:
Donovan, Arthur (1996), *Antoine Lavoisier: Science, Administration, and Revolution*, Cambridge University Press, Cambridge [ISBN 978-0521562188].
Grey, Vivian (1982), *The Chemist Who Lost His Head: The Story of Antoine Lavoisier*, Coward, McCann and Geoghegan, Inc, USA [ISBN 978-0698205598].
Holmes, Frederic Lawrence (1984), *Lavoisier and the Chemistry of Life*, University of Wisconsin Press, Madison, Wisconsin [ISBN 978-0299099800].
Jackson, Joe (2005), *A World on Fire: A Heretic, An Aristocrat And The Race to Discover Oxygen*, Viking, New York [ISBN 978-0670034345].
Smartt Bell, Madison (2005), *Lavoisier in the Year One: The Birth of a New Science in an Age of Revolution*, Atlas Books, W. W. Norton, USA [ISBN 978-0393051551].

Figure Credits

4. Figure 24.1 courtesy: Original drawing by William Alexander. Public domain via Wikimedia Commons.
https://commons.wikimedia.org/wiki/File:Cavendish-walk.jpg (from public domain).
5. Figure 24.2 courtesy: Henry Cavendish [Public domain], via Wikimedia Commons. https://commons.wikimedia.org/wiki/File:Cavendish_hydrogen.jpg (from public domain).
6. Figure 24.3 courtesy: Ellen Sharples (1769 - 1849) [Public domain], via Wikimedia Commons.
https://commons.wikimedia.org/wiki/File:Priestley.jpg (from public domain).
7. Figure 24.4 courtesy: From the Special Collections & Archives Research Center, Oregon State University Libraries; reproduced with permission.

Index

T. Padmanabhan and V. Padmanabhan, *The Dawn of Science*,
https://doi.org/10.1007/978-3-030-17509-2